KV-579-463

Macmillan Maths Topics
General Editor: W E Goss

Algebra

P Whittaker
Head of Mathematics
Ward End Hall School
Birmingham

20.4.79

Macmillan Education

LIVERPOOL INSTITUTE
OF HIGHER EDUCATION
M LIBRARY
WOOLTON ROAD,
LIVERPOOL, L16 8ND

© P. Whittaker 1977

All rights reserved. No part of this
publication may be reproduced or
transmitted, in any form or by any
means, without permission.

First published 1977 by
MACMILLAN EDUCATION LIMITED
Houndmills Basingstoke Hampshire RG21 2XS
and London
Associated companies in New York Dublin
Melbourne Johannesburg and Delhi

Printed in Great Britain by
A. Wheaton and Co., Exeter

British Library Cataloguing in Publication Data
Whittaker, P
Algebra. – (Macmillan maths topics ; 5).
1. Algebra
I. Title II. Series
512.9   QA152.2

ISBN 0–333–19020–3

S. KATHARINE'S
COLLEGE
LIVERPOOL
CLASS :
ACCESS: 20. 4. 79

372.38T SEC
MAC

Contents

L. I. H. E.
THE MARKLAND LIBRARY

Accession No.
~~Stk~~ 28218

Class No. 372.38T
SEC

Catal.

Introduction

This book covers the core of the algebra required by most CSE syllabuses.

Exercises are arranged so that the various aspects of the course are presented in simple stages, enabling the pupil to progress by small steps. Explanations and worked examples, when included, are not intended to be extensive, but merely to provide the pupil with sufficient information to deal with the exercises.

There is no section on graphs, since a comprehensive coverage of graph work is to be found in *Graphs* by R. R. Joy, in the same series.

The final chapter of the book contains a large number of examples taken from past CSE Mathematics papers, and thanks are due to the examining boards listed below for permission to use these questions. Each question is individually credited in the body of the book.

West Midlands Examinations Board (*WMEB*)
East Midlands Regional Examinations Board (*EMREB*)
The West Yorkshire and Lindsey Regional Examining Board (*WYLEB*)
North Western Secondary School Examination Board (*NWSSEB*)
Southern Regional Examinations Board (*SREB*)
Middlesex Regional Examining Board (*MREB*)
Metropolitan Regional Examinations Board (*Met. REB*)
Welsh Joint Education Committee (*WJEC*)
East Anglian Examinations Board (*EAEB*).

1 Basic notions and procedures

This chapter, which covers elementary notions and procedures, has been arranged in such a way as to provide numerous examples for introducing, or revising and reinforcing, ideas and processes, some of which may have been met in earlier school work. For this reason explanations and worked examples have been omitted from the early exercises.

All of the exercises, however, have been graded so that they provide a full coverage of each topic from its simplest form.

The more involved aspects of the chapter are explained and illustrated with worked examples.

1.1 Positive and negative integers

Exercise 1

1. $8+3$	7. $8-6$	13. $8-11$
2. $9+7$	8. $7-5$	14. $7-12$
3. $6+11$	9. $11-4$	15. $11-13$
4. $5+15$	10. $9-3$	16. $9-20$
5. $17+6$	11. $2-0$	17. $2-5$
6. $8+13$	12. $8-5$	18. $8-16$

19. $-8+9$	25. $-8+3$	31. $-8-9$
20. $-7+12$	26. $-7+5$	32. $-5-12$
21. $-3+4$	27. $-3+1$	33. $-6-6$
22. $-6+20$	28. $-15+10$	34. $-3-0$
23. $-3+15$	29. $-40+25$	35. $-15-8$
24. $-6+8$	30. $-6+0$	36. $-11-2$

Exercise 2

1. $9-3$	4. $-6+11$	7. $8+3$
2. $3-9$	5. $-4+4$	8. $-18+10$
3. $-9-3$	6. $0-8$	9. $-5-6$

10. $11 - 15$
11. $12 - 8 + 4$
12. $-6 - 7 + 8$
13. $6 + 7 - 8$
14. $-6 + 7 - 8$
15. $-6 - 7 - 8$
16. $0 - 4 - 5$

17. $5 - 4 + 0$
18. $16 - 9 - 3$
19. $8 - 4 - 3 - 2$
20. $5 + 11 - 6 - 10$
21. $5 - 4 + 3 - 2 + 1$
22. $-5 + 4 - 3 + 2 - 1$
23. $17 - 28 + 6 + 2$

24. $-8 - 4 + 3 - 11$
25. $8 - 6 - 5 + 7 - 2$
26. $100 - 56 + 3 - 9$
27. $-1 - 2 - 3 - 4 - 5$
28. $5 - 4 - 3 - 2 - 1$
29. $17 + 6 - 25 + 8$
30. $8 + 11 - 3 + 6 - 20$

Exercise 3

1. 8×3
2. 9×7
3. 6×11
4. 5×5
5. 4×9
6. 6×12

7. $8 \times (-3)$
8. $9 \times (-7)$
9. $6 \times (-11)$
10. $5 \times (-5)$
11. $4 \times (-9)$
12. $6 \times (-12)$

13. $(-8) \times 3$
14. $(-9) \times 7$
15. $(-6) \times 11$
16. $(-5) \times 5$
17. $(-4) \times 9$
18. $(-6) \times 12$

19. $(-8) \times (-3)$
20. $(-9) \times (-7)$
21. $(-6) \times (-11)$
22. $(-5) \times (-5)$
23. $(-4) \times (-9)$
24. $(-6) \times (-12)$

Exercise 4

1. $3 \times (-2)$
2. 4×8
3. $(-5) \times (-6)$
4. $(-9) \times 8$
5. 4^2
6. $(-4)^2$
7. $(-4) \times 4$
8. $3 \times (-11)$
9. $11 \times (-3)$
10. $(-8) \times 7$

11. 0×4
12. $(-5)^2$
13. $(-5) \times (-5)$
14. $(-10) \times 0$
15. 6×9
16. 7^2
17. $(-2) \times 1$
18. $3 \times (-4)$
19. $(-6) \times (-9)$
20. $(-9) \times (-6)$

21. $1 \times (-1)$
22. $(-1)^2$
23. 4×3
24. $(-4) \times 3$
25. $7 \times (-3)$
26. 8×8
27. $(0)^2$
28. $3 \times (-12)$
29. $(-17) \times (-1)$
30. $(-7) \times 2$

Exercise 5

1. $\dfrac{9}{3}$
2. $\dfrac{8}{16}$
3. $\dfrac{15}{15}$
4. $\dfrac{27}{3}$
5. $\dfrac{12}{36}$

6. $\dfrac{48}{8}$
7. $\dfrac{9}{(-3)}$
8. $\dfrac{8}{(-16)}$
9. $\dfrac{15}{-15}$
10. $\dfrac{27}{-3}$

11. $\dfrac{12}{-36}$
12. $\dfrac{48}{-8}$
13. $\dfrac{-9}{3}$
14. $\dfrac{(-8)}{16}$
15. $\dfrac{(-15)}{15}$

16. $\dfrac{-27}{3}$
17. $\dfrac{-12}{36}$
18. $\dfrac{-48}{8}$
19. $\dfrac{(-9)}{(-3)}$
20. $\dfrac{-8}{-16}$

21. $\dfrac{-15}{-15}$ 22. $\dfrac{(-27)}{-3}$ 23. $\dfrac{-12}{-36}$ 24. $\dfrac{-48}{-8}$

Exercise 6

1. $\dfrac{25}{-5}$ 6. $\dfrac{-5}{-10}$ 11. $\dfrac{9}{9}$ 16. $\dfrac{-10}{-2}$

2. $\dfrac{-100}{-10}$ 7. $\dfrac{6}{-3}$ 12. $\dfrac{-20}{5}$ 17. $\dfrac{-3}{-6}$

3. $\dfrac{56}{7}$ 8. $\dfrac{7}{-3}$ 13. $\dfrac{4}{-1}$ 18. $\dfrac{18}{-4}$

4. $\dfrac{-12}{4}$ 9. $\dfrac{-16}{4}$ 14. $\dfrac{0}{3}$

5. $\dfrac{5}{10}$ 10. $\dfrac{8}{-2}$ 15. $\dfrac{15}{-5}$

Exercise 7

1. $(6)+(4)$ 13. $(6)+(-8)$ 25. $(-6)+4$
2. $(5)+(3)$ 14. $7+(-12)$ 26. $(-10)+3$
3. $(9)+(12)$ 15. $9+(-20)$ 27. $(-2)+1$
4. $(0)+(11)$ 16. $15+(-15)$ 28. $(-6)+(3)$
5. $(7)+(7)$ 17. $(7)+(-10)$ 29. $(-20)+14$
6. $(8)+(1)$ 18. $1+(-3)$ 30. $(-17)+8$
7. $6+(-4)$ 19. $(-6)+8$ 31. $(-6)+(-4)$
8. $(5)+(-3)$ 20. $(-3)+7$ 32. $(-7)+(-12)$
9. $(8)+(-1)$ 21. $(-2)+(10)$ 33. $(-8)+(-1)$
10. $4+(-3)$ 22. $(-10)+15$ 34. $(-15)+(-15)$
11. $7+(-7)$ 23. $(-4)+(6)$ 35. $(-4)+(-7)$
12. $6+(-5)$ 24. $(-4)+5$ 36. $(-1)+(-3)$

Exercise 8

1. $(6)-(4)$ 7. $6-(-4)$ 13. $(-6)-8$
2. $5-(3)$ 8. $(5)-(-3)$ 14. $(-3)-(7)$
3. $(9)-(12)$ 9. $(8)-(-1)$ 15. $(-2)-(+10)$
4. $(0)-11$ 10. $(4)-(-3)$ 16. $(-10)-(+15)$
5. $7-(7)$ 11. $7-(-7)$ 17. $(-4)-(6)$
6. $(8)-1$ 12. $6-(-5)$ 18. $(-4)-5$

19. $(-6)-(-4)$ 21. $(-8)-(-1)$ 23. $(-4)-(-7)$

20. $(-7)-(-12)$ 22. $(-15)-(-15)$ 24. $(-1)-(-3)$

Exercise 9

1. $7+(-8)$
2. $(-3)+6-(-1)$
3. $-20+(-60)$
4. $(15)-(15)$
5. $22-(-11)+(6)$
6. $(-9)+(4)$
7. $(0)-(-1)$
8. $(-2)-(-3)-(-1)$
9. $(-6)+(-4)-(-5)$
10. $7-11+(-3)$

11. $5+(-4)-(3)+(-2)$
12. $5-(-4)-(-3)-(-2)$
13. $0+(-2)+3$
14. $4-(-4)$
15. $(-5)+(-3)+(-1)$
16. $(6)-(4)-(2)$
17. $18-(-3)$
18. $(-18)+(-3)$
19. $4-(-5)+1$
20. $(10)-(10)$

1.2   Collection of terms

Exercise 10

1. $a+a+a+a$
2. $a-a-a-a$
3. $2b+3b$
4. $d+4d-3d$
5. $7h-12h$
6. $x-2x-5x$
7. $10c+6c-4c+3c$
8. $-8e+8f$
9. $-2a-3a+a$
10. $-10y+3y-8y$
11. $4p+3p-9p+7p$
12. $-2f+6f-4f$

13. $a+b+a+b$
14. $a-b+a-b$
15. $-a+b-a+b$
16. $3c-4d+6d-2c$
17. $5x+4y-3x+8y$
18. $-7p+6r+8p-8r$
19. $3a-b+2c-2b+a$
20. $-2x-3y+5x-7y$
21. $10r+6s+5r+2s$
22. $-2a-2b-a-3b$
23. $16f+4g-8f-8g$
24. $s-6t+r-5t+3s-2r$

Exercise 11

Arrange in ascending powers.

1. $2x^2-x^3+5x$
2. $x^4-2x^2-3x^2+2x$
3. $-y+3y^3-2y+y^2-y^3$
4. $7+2x-3x+4x^2$

5. $x^3+2x^3-3x^2+4x-2$
6. t^3+8
7. $m^5+3m-2+6m-2m^5$
8. $x-7x^2+2x^2-x^3+5x$

Arrange in descending powers.

9. $7a^2+8a-a^2-3a^3+2a$
10. $3x^3-2x^2+7x$

11. $a-2a^2+4a+a^3$
12. $2t^2-3t^2+4t+t^2$

13. $4 - 11t - 6t + 3 + 17t$

14. $8x^5 + 3x - x^2 + 4x$

15. $y - 3y^2 - 2y + y^3$

16. $a^2 - 3a + 2a - 4a^3$

1.3 Simple multiplication

Exercise 12

1. $2 \times g$
2. $t \times 4$
3. $a \times b \times c$
4. $2x \times y$
5. $3r \times 4s$
6. $2h \times 9k$
7. $a \times 4b$
8. $2x \times 2y \times 2z$
9. $t \times 4s$
10. $p \times q \times 3r \times 2s$
11. $3ab \times 2c$

12. $pq \times 5r$
13. $3hk \times 4l$
14. $m \times 3n \times 2pq$
15. $8 \times 9r$
16. $4ab \times 2c$
17. $6lm \times 6np$
18. $5rs \times 7$
19. $2bc \times 4ad$
20. $9fg \times 2eh$
21. $a \times a \times a$
22. $2x \times 4x$

23. $3y \times 3y$
24. $t \times 4t \times 2t$
25. $3b \times 5b$
26. $3x \times 2x \times x$
27. $y \times y \times y \times y \times y$
28. $6y \times 2y$
29. $2p \times 3p$
30. $8x \times x \times x$
31. $x \times 6x$
32. $2x \times 2x \times 2x$

1.4 Multiplication and division (involving indices)

Exercise 13

1. $x^2 \times x^3$
2. $y^4 \times y^2$
3. $a^3 \times a$
4. $b \times b^5$
5. $x^3 \times x^4$
6. $y^7 \times y^3$

7. $r^2 \times rs^3$
8. $x \times x^2 y^2$
9. $x^4 \times xy$
10. $y^3 \times y^2 z^2$
11. $p \times pr$
12. $x^2 \times x^2 y$

13. $a^2 b \times a^3$
14. $xy^5 \times x^2$
15. $pr \times r^4$
16. $c^2 d^2 \times c^2$
17. $c^2 d^2 \times d^2$
18. $x^2 y \times y$

19. $xy^3 \times x^2 y$
20. $p^2 r^3 \times pr$
21. $mn^5 \times m^3 n^2$
22. $m^2 n^2 \times m^2 n^2$
23. $yz \times yz$
24. $b^2 c \times b^3 c^3$

Exercise 14

1. $2x^2 \times 3x^3$
2. $4y^4 \times 2y^2$
3. $5a^3 \times 4a$
4. $6b \times 6b^5$

5. $2x^3 \times 5x^4$
6. $8y^7 \times y^3$
7. $2x^2 y \times 3yz$
8. $4t \times 3r^2 t^2$

9. $xyz \times 2y^2$
10. $2x \times 2y^2 \times 3z$
11. $ab^2 \times 3a^2 b^2 \times 2c$
12. $x^2 \times x^2 \times x^2$

Exercise 15

1. $\dfrac{x^5}{x^2}$
2. $\dfrac{y^9}{y^3}$
3. $\dfrac{a^2}{a}$
4. $\dfrac{y^7}{y^6}$
5. $\dfrac{b^2}{b^2}$
6. $\dfrac{x^5}{x}$

7. $\dfrac{x^2}{x^5}$

13. $\dfrac{x^4 y^3}{x^2 y}$

19. $\dfrac{xy^2}{x^3 y^3}$

25. $\dfrac{x^6 y^2}{xy^4}$

8. $\dfrac{y^3}{y^9}$

14. $\dfrac{a^2 b^4}{ab^2}$

20. $\dfrac{pqr}{p^2 q^2 r^2}$

26. $\dfrac{xy}{xz}$

9. $\dfrac{a}{a^2}$

15. $\dfrac{at^3}{t^2}$

21. $\dfrac{t^2}{st^4}$

27. $\dfrac{2t^2}{4s}$

10. $\dfrac{y^6}{y^7}$

16. $\dfrac{x^6 y^2 z^3}{x^4 yz}$

22. $\dfrac{xy}{x^2 y^3}$

28. $\dfrac{10st}{2s^2}$

11. $\dfrac{b^5}{b^5}$

17. $\dfrac{st^3}{t^3}$

23. $\dfrac{xy}{xy}$

29. $\dfrac{klm}{2l}$

12. $\dfrac{x}{x^5}$

18. $\dfrac{a^2 bc^3}{a^2 c^2}$

24. $\dfrac{p^2 q}{p^4 q}$

30. $\dfrac{x^2 y}{xy^2}$

Exercise 16

1. $\dfrac{6x^4}{3x^2}$

6. $\dfrac{2x^6}{7x^5}$

11. $\dfrac{8ab^2}{12b^2}$

16. $\dfrac{21ac^3}{14bc^2}$

2. $\dfrac{2x^6}{10x}$

7. $\dfrac{2x^2 y}{4xy^3}$

12. $\dfrac{x^2 y^3}{5xy}$

17. $\dfrac{11ab}{11ab}$

3. $\dfrac{3y^4}{y^4}$

8. $\dfrac{15abc^2}{3a^2 b}$

13. $\dfrac{a^2 bc}{ab^2 c}$

18. $\dfrac{33xy^2}{44y^2 z}$

4. $\dfrac{12a^2}{2a}$

9. $\dfrac{18x^2 y^2}{4xy^4}$

14. $\dfrac{3xyz}{x^2 y^2}$

19. $\dfrac{24ab}{8b^2 c}$

5. $\dfrac{12a^2}{2a^4}$

10. $\dfrac{3a}{12b}$

15. $\dfrac{m^2 n}{mn^2}$

20. $\dfrac{15x}{18x^2 y^2}$

1.5 Substitution

Exercise 17a

If $a = 4$, find the value of

1. $3a$
2. $a + 3$
3. $20 - 3a$

4. $5a + 6$
5. $2a - 20$
6. $7a + 2a$

7. $4a - 6 + a$
8. $15 + 3a - 2a$
9. $9a - 7$

10. $6a - 2a$
11. $17 + a$
12. $-28a + 7a$

Exercise 17b

If $a = (-3)$, find the value of

1. $2a$
2. $a + 5$
3. $6 + 2a$

4. $6 - 2a$
5. $3a + 15$
6. $5a + 2a$

7. $5a - 2a$
8. a^2
9. $2 + a + 5$

10. $2a - 1$
11. $a + 1 - 2a$
12. $a + a - a$

Exercise 18a

If $x = 2$ and $y = 3$, evaluate

1. $x + y$
2. $y - x$
3. $x - y$

4. xy
5. yx
6. $2y + x$

7. $3x - 2y$
8. $2xy + 3$
9. $5 - 2x + 3y$

10. $10x - xy$
11. x^2
12. $y^2 - 4x$

Exercise 18b

If $x = (-2)$ and $y = 5$, evaluate

1. $x + y$
2. $x - y$
3. $y - x$

4. xy
5. $2y + x$
6. $2x - y$

7. $x + xy$
8. $x^2 + y$
9. $2x + 2y$

10. $5 - 3x$
11. $10x + 5y$
12. $xy + 10$

Exercise 19a

If $a = 2$, $b = 3$, $c = 1$, $d = 0$, find the value of

1. ab
2. $2cd$
3. $5b^2$

4. $3ac^2$
5. $abcd$
6. $(3bc)^2$

7. a^2b
8. ab^2
9. c^2d^2

10. a^2b^2
11. a^3
12. $2a^3$

Exercise 19b

If $a = 2$, $b = (-1)$, $c = (-4)$, $d = 0$, find the value of

1. ab
2. $5b$
3. $c + b$

4. $c - b$
5. bd
6. c^2

7. abc
8. $c^2 - b^2$
9. $2a + c$

10. $2a - c$
11. bc
12. $abcd$

Exercise 20a

If $r = 2$, $s = 3$, $t = 5$, $x = 6$, $y = 10$, evaluate

1. $\dfrac{y}{r}$
2. $\dfrac{x}{r}$
3. $\dfrac{x}{s}$

4. $\dfrac{s}{x}$
5. $\dfrac{2y}{t}$
6. $\dfrac{rs}{x}$

7. $\dfrac{y}{tx}$
8. $\dfrac{r+s}{t}$
9. $\dfrac{2t}{4y}$

10. $\dfrac{s^2}{x}$
11. $\dfrac{rst}{2y}$
12. $\dfrac{r^2s^2}{x^2}$

11

Exercise 20b

If $r = (-2)$, $s = (-3)$, $t = 5$, $x = 6$, $y = (-10)$, evaluate

1. $\dfrac{y}{r}$ 4. $\dfrac{s}{r}$ 7. $\dfrac{rs}{x}$ 10. $\dfrac{r-s}{t}$

2. $\dfrac{y}{t}$ 5. $\dfrac{r}{s}$ 8. $\dfrac{3y}{st}$ 11. $\dfrac{s^2}{s}$

3. $\dfrac{x}{r}$ 6. $\dfrac{tx}{y}$ 9. $\dfrac{r+s}{t}$ 12. $\dfrac{rsy}{tx}$

Exercise 21

Copy and complete the following table, using the values given.

x	y	$x+y$	$x-y$	xy	$\dfrac{x}{y}$
5	3				
0	4				
1	1				
-4	3				
2	-1				
-2	-3				

Exercise 22

Find the value of $3x - 4$ when

1. $x = 2$ 3. $x = (-2)$ 5. $x = 1$ 7. $x = 2\frac{1}{2}$ 9. $x = (-3)$
2. $x = 0$ 4. $x = 5$ 6. $x = 4$ 8. $x = (-1)$ 10. $x = \frac{4}{3}$

Evaluate $2xyz$ when

11. $x = 2$, $y = 1$, $z = 3$ 15. $x = (-3)$, $y = 1$,
12. $x = (-3)$, $y = 0$, $z = 5$ $z = (-5)$
13. $x = (-1)$, $y = (-1)$, $z = (-1)$ 16. $x = 3$, $y = \frac{1}{3}$, $z = (-\frac{1}{2})$
14. $x = \frac{1}{2}$, $y = 2$, $z = (-5)$

Calculate the value of $x^2 - 3x + 2$ when

17. $x = 0$ 19. $x = (-1)$ 21. $x = (-6)$ 23. $x = 2$
18. $x = 1$ 20. $x = 3$ 22. $x = 5$ 24. $x = (-2)$

12

Find the value of $x^2 + 2x - 3$ when

25. $x = 5$ 27. $x = (-3)$ 29. $x = 0$ 31. $x = 4$
26. $x = (-1)$ 28. $x = 2$ 30. $x = 1$ 32. $x = (-5)$

1.6 Brackets I – expansions

Exercise 23

Remove the brackets and simplify.

1. $11 + (6 + 3)$
2. $3 + (4 + 5)$
3. $2x + (3x + 2)$
4. $4 + (5 + 2x)$
5. $(2x + 3) + 4x$
6. $(4y + 3y) + 2y$

7. $11 + (6 - 3)$
8. $3 + (4 - 5)$
9. $2x + (3x - 2)$
10. $4 + (5 - 2x)$
11. $(2x - 3) + 4x$
12. $(4y - 3y) + 2y$

13. $11 - (6 + 3)$
14. $3 - (4 + 5)$
15. $2x - (3x + 2)$
16. $4 - (5 + 2x)$
17. $(2x + 3) - 4x$
18. $(4y + 3y) - 2y$

19. $11 - (6 - 3)$
20. $3 - (4 - 5)$
21. $2x - (3x - 2)$
22. $4 - (5 - 2x)$
23. $(2x - 3) - 4x$
24. $(4y - 3y) - 2y$

Exercise 24

Remove the brackets and simplify.

1. $14 - (3 - 6)$
2. $(7 - 2) + (3 + 4)$
3. $a + (2a - b)$
4. $(b + c) - (b - c)$
5. $(a - b + c) - b$
6. $(2a + b) + (2a - b)$
7. $-(x - 2y + z) + (x + z)$
8. $2a + 3b - (4b - 6a)$
9. $(x - y) - z$

10. $-2x + (y - 2x)$
11. $(3a + b) - (2a + b)$
12. $(a + b) + (a - b) - (b + a)$
13. $5 - a + (a + 5)$
14. $(2x + 6) - x + 3$
15. $4a - (3 - 2a) + 2$
16. $1 + (8 - x)$
17. $(a - b) + (b - c) - (c - a)$
18. $2p + (3q - 4r)$

Exercise 25a

Expand the brackets.

1. $3(x + y)$
2. $7(s + t)$
3. $4(m + 6)$

4. $2(3 + k)$
5. $3(2a + 5b)$
6. $\frac{1}{2}(6x + 10)$

7. $5(a + 2b + 6)$
8. $6(x + y + \frac{1}{4})$

13

9. $2(a-b)$
10. $3(x-y)$
11. $5(t-2)$
12. $3(4-d)$
13. $\frac{1}{4}(12x-4y)$
14. $7(2a-3b)$
15. $4(x-2y-3)$
16. $6(2a-b+2)$

17. $-3(x+y)$
18. $-7(s+t)$
19. $-4(m+6)$
20. $-2(3+k)$
21. $-3(2a+5b)$
22. $-\frac{1}{2}(4x+9)$
23. $-5(a+2b+6)$
24. $-6(x+y+\frac{3}{4})$

25. $-2(a-b)$
26. $-3(x-y)$
27. $-5(t-2)$
28. $-3(4-d)$

29. $-\frac{1}{4}(8x-2y)$
30. $-7(2a-3b)$
31. $-4(x-2y-3)$
32. $-6(2a-b+2)$

Exercise 25b

Expand the brackets.

1. $a(x+y)$
2. $s(s+t)$
3. $k(m+6)$
4. $k(3+k)$
5. $x(2a+5b)$

6. $b(a-b)$
7. $a(x-y)$
8. $t(t-2)$
9. $e(4-d)$
10. $x(12x-4y)$

11. $-a(x+y)$
12. $-s(s+t)$
13. $-k(m+6)$
14. $-k(3+k)$
15. $-x(2a+5b)$

16. $-b(a-b)$
17. $-a(x-y)$
18. $-t(t-2)$
19. $-e(4-d)$
20. $-x(12x-4y)$

Exercise 26

Expand the brackets.

1. $2a(3a-4b)$
2. $-x(y-x)$
3. $2xy(x+y)$
4. $x^2(3-x)$

5. $-2y(3+x-y)$
6. $2a(a^2+a)+a$
7. $3x-2(y-4)$
8. $xyz(x+y+z)$

9. $-4x(x-2y+3z)$
10. $4a(b-c)-2a$
11. $x(1-t)$
12. $pq(p^2-q^2)$

Exercise 27

Expand the brackets and simplify by collecting like terms.

1. $3(x+2)+2(x+5)$
2. $6(x+1)+5(x-1)$
3. $4(4-x)+3(2x+6)$
4. $7(3+2x)+2(x-7)$

5. $3(x-1)+(2x+4)$
6. $2(2x-6)-3(x+3)$
7. $3(4+3x)-(6x+5)$
8. $5(x-1)-2(x+1)$

14

9. $2(3-x)-4(x-3)$ 11. $3(4x-2)+2(3x+4)-4(x-1)$
10. $5(x+1)-4(x+1)$ 12. $5(3-2x)-(5x+4)+2(4x-3)$

Expanding 'double brackets'

Consider:

1. $\underline{(a+b)}\ (c+d)$ means $\underline{a}(c+d)\underline{+b}(c+d)$

                               i.e. $ac+ad+bc+bd.$

2. $\underline{(x+2)}\ (x+3)$ means $\underline{x}(x+3)\underline{+2}(x+3)$

                               i.e. $x^2+3x+2x+6.$

Collect like terms   i.e. $x^2+5x+6.$

Exercise 28

Expand the brackets, and simplify where possible.

1. $(a+b)(m+n)$ 8. $(2+y)(c+d)$ 15. $(x+1)(x+2)$
2. $(c+d)(x+y)$ 9. $(a+8)(a+2)$ 16. $(a+6)(a+3)$
3. $(a+2)(a+3)$ 10. $(x+5)(x+5)$ 17. $(y+5)(y+4)$
4. $(b+4)(b+1)$ 11. $(x+a)(y+a)$ 18. $(2+y)(3+y)$
5. $(x+2)(x+4)$ 12. $(t+7)(t+8)$ 19. $(x+4)(x+8)$
6. $(4+a)(3+a)$ 13. $(3+x)(y+z)$ 20. $(x+2)(x+7)$
7. $(a+b)(x+2)$ 14. $(5+p)(5+p)$

Consider:

3. $\underline{(a+b)}\ (c-d)$ means $\underline{a}(c-d)\underline{+b}(c-d)$

                               i.e. $ac-ad+bc-bd.$

4. $\underline{(x+2)}\ (x-3)$ means $\underline{x}(x-3)\underline{+2}(x-3)$

                               i.e. $x^2-3x+2x-6.$

Collect like terms   i.e. $x^2-x-6.$

5. $\underline{(x+3)}\ (x-1)$ means $\underline{x}\ (x-1)+3(x-1)$

                               i.e. $x^2-x+3x-3.$

Collect like terms   i.e. $x^2+2x-3.$

Exercise 29

Expand the brackets and simplify where possible.

1. $(a+b)(x-y)$
2. $(x+3)(x-4)$
3. $(x+4)(x-3)$
4. $(2+x)(3-x)$
5. $(a+5)(a-6)$
6. $(x+5)(x-1)$

7. $(x+3)(x-3)$
8. $(a+2)(a-8)$
9. $(t+1)(t-2)$
10. $(4+y)(3-y)$
11. $(x+5)(x-7)$
12. $(x+6)(x-2)$

13. $(x+7)(x-5)$
14. $(x+5)(x-5)$
15. $(x+10)(x-3)$
16. $(x+y)(x-y)$

Consider:

6. $\underline{(a-b)}(c+d)$ means $\underline{a(c+d)}-b(c+d)$

   i.e. $ac+ad-bc-bd.$

7. $\underline{(x-2)}\,(x+3)$ means $\underline{x}\,(x+3)-2(x+3)$

   i.e. $x^2+3x-2x-6.$

Collect like terms i.e. $x^2+x-6.$

8. $\underline{(x-3)}\,(x+1)$ means $x(x+1)-3(x+1)$

   i.e. $x^2+x-3x-3.$

Collect like terms i.e. $x^2-2x-3.$

Exercise 30

Expand the brackets and simplify where possible.

1. $(a-b)(x+y)$
2. $(x-3)(x+4)$
3. $(x-4)(x+3)$
4. $(1-x)(4+x)$
5. $(a-3)(a+5)$
6. $(x-6)(x+3)$

7. $(x-2)(x+2)$
8. $(a-5)(b+7)$
9. $(t-2)(t+3)$
10. $(4-y)(2+y)$
11. $(x-6)(x+10)$
12. $(x-3)(x+1)$

13. $(x-2)(x+1)$
14. $(x-6)(x+6)$
15. $(x-12)(x+2)$
16. $(x-5)(x+7)$

Consider:

9. $\underline{(a-b)}\,(c-d)$ means $\underline{a(c-d)}-b(c-d)$

   i.e. $ac-ad-bc+bd.$

10. $\underline{(x-2)}\,(x-3)$ means $\underline{x(x-3)}-2(x-3)$

    i.e. $x^2-3x-2x+6.$

Collect like terms i.e. $x^2-5x+6.$

16

Exercise 31

Expand the brackets and simplify where possible.

1. $(x-y)(a-b)$
2. $(x-4)(x-5)$
3. $(x-1)(x-3)$
4. $(a-3)(a-2)$
5. $(4-y)(3-y)$
6. $(a-6)(a-6)$

7. $(2-t)(t-4)$
8. $(x-1)(x-2)$
9. $(x-7)(x-3)$
10. $(x-a)(a-x)$
11. $(y-5)(y-6)$
12. $(x-1)(x-1)$

13. $(x-10)(x-2)$
14. $(x-3)(x-4)$
15. $(x-7)(x-9)$
16. $(x-y)(x-y)$

Perfect squares

$(x+5)^2$ means $(x+5)(x+5)$, i.e. $x(x+5)+5(x+5)$

i.e. $x^2+5x+5x+25$

i.e. $x^2+10x+25$.

$(x-3)^2$ means $(x-3)(x-3)$, i.e. $x(x-3)-3(x-3)$

i.e. $x^2-3x-3x+9$

i.e. x^2-6x+9.

Exercise 32

Expand the following.

1. $(x+4)^2$
2. $(x+1)^2$
3. $(x+6)^2$

4. $(x-2)^2$
5. $(x-5)^2$
6. $(x-a)^2$

7. $(x+8)^2$
8. $(x-7)^2$
9. $(a+x)^2$

10. $(4-t)^2$
11. $(x+10)^2$
12. $(x-9)^2$

Special case (difference of two squares)

$(a-b)(a+b)$ expands to $a^2+ab-ab-b^2$ i.e. a^2-b^2.
$(x-3)(x+3)$ expands to $x^2+3x-3x-9$ i.e. x^2-9.
$(x+5)(x-5)$ expands to $x^2-5x+5x-25$ i.e. x^2-25.

Hence when the brackets to be expanded are the same, except for *opposite signs*, there is *no middle term* in the answer.

Note that the two terms in the answer are both *perfect squares*. This special case is referred to as the DIFFERENCE OF TWO SQUARES. It will be met again in the next section on factorisation!

A—B

Exercise 33

In this exercise you are recommended to show full working for the first six questions, and then write down the answers to the other questions directly.

1. $(x-y)(x+y)$
2. $(x+4)(x-4)$
3. $(x-1)(x+1)$
4. $(m+n)(m-n)$

5. $(x-10)(x+10)$
6. $(7+x)(7-x)$
7. $(x+2)(x-2)$
8. $(a-x)(a+x)$

9. $(x-6)(x+6)$
10. $(4+x)(4-x)$
11. $(x+7)(x-7)$
12. $(1-x)(1+x)$

Harder expansions

Consider:

1. $(5x-3)(3x+4)$ means $5x(3x+4)-3(3x+4)$
    i.e. $15x^2+20x-9x-12$
    i.e. $15x^2+11x-12.$

2. $(2x-3y)^2$ means $(2x-3y)(2x-3y)$
    i.e. $2x(2x-3y)-3y(2x-3y)$
    i.e. $4x^2-6xy-6xy+9y^2$
    i.e. $4x^2-12xy+9y^2.$

Exercise 34

Expand the following.

1. $(4x+3y)(x+2y)$
2. $(2x+1)(6x+3)$
3. $(2x+5)^2$
4. $(5x-2)(2x+3)$
5. $(3x-7)(2x+1)$
6. $(x-9)(4x+5)$

7. $(3+2x)(4-5x)$
8. $(y+3x)(2y-x)$
9. $(3+5y)(2y-1)$
10. $(3x-4y)^2$
11. $(6x-1)(3x-2)$
12. $(2x-y)(3y-2x)$

Exercise 35

Miscellaneous examples.

1. $(a+8)(a-6)$
2. $(2x+3)(x+4)$
3. $(x-y)(x+y)$
4. $(3x-2)(4x-1)$
5. $(2-y)(y+5)$
6. $(6x+1)(6x+1)$
7. $(5-x)^2$
8. $(p-1)(p-2)$

9. $(a-6)(a+3)$
10. $(2x+1)(3x+4)$
11. $(m+12)^2$
12. $(2x+3)^2$
13. $(2x+3y)(3x+2y)$
14. $(10x-10)(10x+10)$
15. $(4x+8)(3x-9)$
16. $(a-3b)(a-2b)$

18

1.7 Brackets II–factorisations

Factorisation is the reverse process to removal (expansion) of brackets.

(a) Common factors

Consider:

1. $3(a+2b) = 3a+6b$. In reverse $3a+6b = 3(a+2b)$.
   The COMMON FACTOR is 3.
2. $x(2x+y) = 2x^2+xy$. In reverse $2x^2+xy = x(2x+y)$.
   The common factor is x.
3. $3ab(c-d) = 3abc-3abd$. In reverse $3abc-3abd = 3ab(c-d)$.
   The common factor is $3ab$.

Exercise 36

Complete the following statements, in which the common factor has been selected for you.

1. $2x+8y = 2($ $)$
2. $6r+3s+9t = 3($ $)$
3. $20x-8y = 4($ $)$
4. $x^2+xy = x($ $)$
5. $4bc-3cd = c($ $)$

6. $5x^2-10xy = 5x($ $)$
7. $xy+x^2y = xy($ $)$
8. $5ab-5b^2+15bc = 5b($ $)$
9. $-12a-4b = -4($ $)$
10. $-2x+10y-4z = -2($ $)$

11. $4x^2-2xy = 2x($ $)$
12. $4xy+12xz = 4x($ $)$
13. $9d-45e = 9($ $)$
14. $21ab+63a = 21a($ $)$
15. $x^2y+xy^2+xy = xy($ $)$

Exercise 37

Factorise.

1. $5a-10b$
2. $6x+3y-12z$
3. $7m+56n$
4. $14r-8s-12t$
5. $3x+6y+6z$
6. $24x-32y+16z$
7. $ax+ay$
8. $cz-dz$
9. x^2+xy
10. $7pq-9pr$

11. $3xy+2y^2-7y$
12. $abc-2bc+4b^2$
13. $3ab+9bc$
14. $4x^2-6xy$
15. $28cd+7d^2$
16. $3x^2y^2-6xyz$
17. $x^2yz+xy^2z-xyz^2$
18. $5bc-25ab^2$
19. $12xy+8x^2$
20. x^4-x^2

(b) Grouping in pairs

Consider:

1. $(a+b)(c+d)$ becomes $a(c+d)+b(c+d)$
i.e. $ac+ad+bc+bd$ (expansion).
In reverse $ac+ad+bc+bd$, grouping in pairs, we have
$$ac+ad+bc+bd, \qquad \text{i.e.} \quad a(c+d)+b(c+d).$$
Note: $(c+d)$ is a common factor $\quad$ i.e. $\quad (a+b)(c+d)$.

2. $3m+3n-am-an$ becomes $\qquad 3(m+n)-a(m+n)$.
The common factor is $(m+n)$ $\qquad$ i.e. $\quad (3-a)(m+n)$.

3. $ax-ay+2x-2y$ becomes $\qquad a(x-y)+2(x-y)$
The common factor is $(x-y)$ $\qquad$ i.e. $\quad (a+2)(x-y)$.

4. $5x+5y+ax-ay$ becomes $\qquad 5(x+y)+a(x-y)$
There is no common factor.

5. $2a-2b-ax+bx$ becomes $\qquad 2(a-b)-x(a-b)$.
The common factor is $(a-b)$ $\qquad$ i.e. $\quad (2-x)(a-b)$.

Exercise 38

Complete the factorisation; the first stage has been done for you.

1. $3(x-y)+a(x-y)$
2. $x(a+b)+y(a+b)$
3. $x(x-2)+y(x-2)$
4. $a(2x+1)-b(2x+1)$
5. $2x(a-5)-5y(a-5)$
6. $p(x-2)-q(x+2)$
7. $4(x^2-1)+y(x^2-1)$
8. $a(b+c)-d(b+c)$
9. $x(3-b)-y(3-b)$
10. $2x(x+1)+y(x+1)$

Can you work out the *original* expressions to which the factors you have just found, apply?

Exercise 39

Factorise.

1. $3x+3y+ax+ay$
2. $ab+4b+5a+20$
3. $ax+ay+2x+2y$
4. $a^2+2a+ab+2b$
5. $tx+3x+3t+9$
6. $5x+5y-bx-by$
7. $ax+3x-6a-18$
8. $4x+4y-tx-ty$
9. $m^2+2m-mn-2n$
10. $2y+xy-2x-x^2$

11. $2x-2y+ax-ay$
12. $xy-4y+5x-20$
13. $7x-7y+x^2-xy$
14. $16-4x+4y-xy$
15. $ax-3x+3a-9$
16. $4x-4y-ax+ay$
17. $3ax-a-6x^2+2x$
18. $4x-4y-tx+ty$
19. $ax-ay-x+y$
20. $5y-xy-5x+x^2$

Exercise 40

In this exercise you will need to re-arrange each expression before you can factorise. If there are no factors, say so.

1. $3a+2x+6+ax$ 5. $5x+4y+ax+by$
2. $xy+2y+2x+y^2$ 6. $-3y+x^2-3x+xy$
3. $ab-cd+ac-bd$ 7. $6+2xy+3x+4y$
4. $ax+14-2a-7x$ 8. $2y+ax-ay-2x$

(c) Factorisation of trinomials ('double brackets')

Consider:

$(x+2)(x+3)$ expands to $x^2+3x+2x+6$
i.e. x^2+5x+6.

The expression x^2+5x+6 is called a TRINOMIAL, because it contains three terms. In order to factorise a trinomial, we need to think very carefully about the expansion of 'double brackets'.

Consider this again:

$(x+2)(x+3)$ becomes $x^2+\underbrace{3x+2x}+6$

$$\downarrow \qquad \downarrow \qquad \downarrow$$

i.e. $x^2+ \quad 5x \quad +6$.

Note that   (i) the first term x^2 of the answer is produced directly from the expansion

(ii) the last term 6 of the answer is produced directly from the expansion

(iii) the middle term $5x$ is a combination of the remaining two parts of the expansion.

Example 1

Factorise x^2+5x+6.
The first term, x^2, could only have been produced by $x \times x$, so we can place these terms inside the brackets immediately, giving

$(x \qquad)(x \qquad)$.

The last term, 6, could have been produced in four ways:

(i) $+6 \times +1 = +6$
(ii) $-6 \times -1 = +6$
(iii) $+3 \times +2 = +6$
(iv) $-3 \times -2 = +6$.

We now have to choose which of these four possibilities combines (by addition) to result in the middle term of $\underline{+5}$ (i.e. $+5x$).

(i) $+6$ and $+1 = +7$
(ii) -6 and $-1 = -7$
(iii) $+3$ and $+2 = +5$
(iv) -3 and $-2 = -5.$

Only possibility (iii) is suitable, i.e. $+3$ and $+2$.
We can now complete the brackets

$(x+3)(x+2)$ are the factors of x^2+5x+6.

Note Try the other possibilities, i.e. $(x+6)(x+1)$
$$(x-6)(x-1)$$
$$(x-3)(x-2)$$

and you will find that none result in x^2+5x+6.

Example 2

Factorise $x^2-8x+16$.

$$x^2 \to x \times x \to (x \qquad)(x \qquad)$$

$$+16 \to +16 \times +1$$
$$-16 \times -1$$
$$+8 \times +2$$
$$-8 \times -2$$
$$+4 \times +4$$

$$\boxed{-4 \times -4}$$

Only -4 and -4 combine to give -8, the middle term, so $(x-4)(x-4)$ are the required factors.

Example 3

Factorise x^2+2x-8.

$$x^2 \to x \times x \to (x \qquad)(x \qquad)$$

$$-8 \to +8 \times -1$$
$$-8 \times +1$$
$$\boxed{+4 \times -2}$$
$$-4 \times +2$$

Only $+4$ and -2 combine to give $+2$, the middle term, so $(x+4)(x-2)$ are the required factors.

Example 4

Factorise $x^2 - 5x - 14$.

$$x^2 \rightarrow x \times x \rightarrow (x \quad)(x \quad)$$

$$-14 \rightarrow -14 \times +1$$
$$+14 \times -1$$
$$\boxed{-7 \times +2}$$
$$+7 \times -2$$

Only -7 and $+2$ combine to give -5, the middle term, so $(x-7)(x+2)$ are the required factors.

Exercise 41

Factorise.

1. $x^2 + 5x + 4$
2. $x^2 + 7x + 12$
3. $x^2 + 4x + 3$
4. $x^2 + 9x + 8$

5. $x^2 + 7x + 6$
6. $x^2 + 9x + 14$
7. $y^2 + 6y + 5$
8. $a^2 + 18a + 32$

9. $x^2 + 13x + 22$
10. $t^2 + 8t + 12$
11. $x^2 + 2x + 1$
12. $x^2 + 13x + 36$

Exercise 42

Factorise.

1. $x^2 - 5x + 6$
2. $x^2 - 6x + 5$
3. $x^2 - 7x + 10$
4. $x^2 - 10x + 24$

5. $a^2 - 10a + 25$
6. $y^2 - 4y + 4$
7. $x^2 - 8x + 7$
8. $x^2 - 8x + 12$

9. $x^2 - 11x + 30$
10. $x^2 - 20x + 36$
11. $t^2 - 29t + 100$
12. $x^2 - 9x + 20$

Exercise 43

Factorise.

1. $x^2 + x - 6$
2. $x^2 - x - 6$
3. $x^2 + 9x - 10$
4. $x^2 + 6x - 7$

5. $x^2 + 2x - 3$
6. $x^2 - x - 12$
7. $t^2 - 8t - 9$
8. $x^2 + 5x - 24$

9. $y^2 - 2y - 24$
10. $x^2 + x - 2$
11. $x^2 - x - 30$
12. $x^2 + 3x - 18$

Perfect squares

When factorising trinomials the result will occasionally be identical brackets, for example $(x+5)(x+5)$. In such cases the answer may be written in the somewhat shorter form $(x+5)^2$, in other words as a PERFECT SQUARE.

Example

Factorise x^2-6x+9.

$$x^2 \to x \times x \to (x \quad)(x \quad)$$

$$+9 \to +9 \times +1$$
$$-9 \times -1$$
$$+3 \times +3$$

$$\boxed{-3 \times -3}$$

Only -3 and -3 combine to give -6, the middle term, so we now have $(x-3)(x-3)$, or more briefly $(x-3)^2$.

Exercise 44

Factorise, and write the result as a perfect square.

1. x^2+6x+9
2. $x^2-20x+100$
3. $x^2+2xy+y^2$
4. $a^2-12a+36$

5. y^2-2y+1
6. $4+4x+x^2$
7. $x^2+14x+49$
8. $x^2-14x+49$

9. $t^2+2at+a^2$
10. $x^2+10x+25$
11. $x^2-18x+81$
12. $4x^2+12x+9$

Difference of two squares

We have already seen that when two brackets are the same, but of opposite sign, then there is no middle term in the resulting expansion. Hence $(x-3)(x+3)$ expands to $x^2+3x-3x-9$, i.e. x^2-9.
More briefly $(x+5)(x-5)$ expands to x^2-5^2, i.e. x^2-25.

Clearly, therefore, in reverse, if we have two terms, both of which are perfect squares, and one is being subtracted from the other, then we may factorise very rapidly: so that

x^2-36 becomes $(x-6)(x+6)$ or $(x+6)(x-6)$
a^2-b^2 becomes $(a-b)(a+b)$
$4x^2-100$ becomes $(2x-10)(2x+10)$.

Remember that this is the DIFFERENCE OF TWO SQUARES, and the procedure does not apply to cases which are the sum of two squares, such as x^2+36!

Exercise 45

Factorise the following; each expression is the difference of two squares.

1. $x^2 - 4$
2. $x^2 - 81$
3. $x^2 - 64$
4. $x^2 - 49$
5. $x^2 - 144$
6. $x^2 - 1$
7. $x^2 - y^2$
8. $x^2 - 9$
9. $16 - x^2$
10. $x^2 - 4y^2$
11. $9x^2 - 25$
12. $4p^2 - 9q^2$
13. $x^2 - 169$
14. $x^2y^2 - 4$
15. $225 - 4x^2$

Harder factorisations

Consider:

1. Factorise $3x^2 + 13x + 4$.
Starting with $3x^2$, we may say $(3x \quad)(x \quad)$.
The last term $+4$ could have been produced by $+4 \times +1$
$$-4 \times -1$$
$$+2 \times +2$$
$$-2 \times -2.$$
None of the four possibilities, when combined *directly*, will give the middle term. This is because the 3 already in the first bracket affects the outcome.

We must try each possibility in turn, until we find the one that works. Hence, we have

$(3x+4)(x+1)$, $(3x-4)(x-1)$, $(3x+2)(x+2)$
$(3x+1)(x+4)$, $(3x-1)(x-4)$, $(3x-2)(x-2)$.

Only $(3x+1)(x+4)$ expands to $3x^2 + 13x + 4$.

2. Factorise $5x^2 + 2x - 3$.
Starting with $5x^2$, we have $(5x \quad)(x \quad)$.
The last term -3 could have been produced by $+3 \times -1$
$$-3 \times +1.$$
Trying each possibility, the only combination which works is
$(5x-3)(x+1)$.

Exercise 46

Factorise.

1. $2x^2 + 5x + 2$
2. $3x^2 - 13x + 4$
3. $2x^2 - x - 3$
4. $2x^2 + 3x - 5$
5. $5x^2 - 17x + 6$
6. $3x^2 + 11x + 6$
7. $3x^2 - 7x + 4$
8. $5 - 24x - 5x^2$

Exercise 47

Miscellaneous examples.
Factorise.

1. $a^2 + 8a - 9$
2. $x^2 + 8x + 16$
3. $25 - x^2$
4. $3x + 3y + ax + ay$
5. $x^2 - 7x + 12$
6. $y^2 - 10y - 11$
7. $4a^2 - b^2$
8. $x^2 - 2x + 1$
9. $t^2 - t + tx - x$
10. $x^2 + 13x + 40$

11. $12 + 8x + x^2$
12. $24 - 14y + y^2$
13. $400 - t^2$
14. $x^2 - 3x - 10$
15. $x^2 - 16x + 64$
16. $9 + 6t + t^2$
17. $4 + 2x - 2y - xy$
18. $x^2 - 121$
19. $2x^2 + 7x + 6$
20. $2x^2 + 6x - 8$

1.8 Fractions

Algebraic fractions obey the same rules as ordinary arithmetic fractions, and if you have a reasonable grasp of the processes of ordinary fractions you will find that the procedures are merely being extended to incorporate letters.

(a) Equivalent fractions (multiplying up/cancelling down)

Consider:

$$\frac{2}{3} = \frac{6}{9}$$ (multiply numerator and denominator by 3)

$$\frac{2a}{5} = \frac{12a}{30}$$ (multiply numerator and denominator by 6)

$$\frac{5}{x} = \frac{10}{2x}$$ (multiply numerator and denominator by 2)

$$\frac{a}{b} = \frac{a^2}{ab}$$ (multiply numerator and denominator by a)

$$\frac{2x}{3y} = \frac{10xy}{15y^2}$$ (multiply numerator and denominator by $5y$).

Further:

$$\frac{12}{24} = \frac{1}{2}$$ (cancel (divide) numerator and denominator by 12)

26

$$\frac{3x}{21} = \frac{x}{7} \qquad \text{(cancel numerator and denominator by 3)}$$

$$\frac{5ab}{6a} = \frac{5b}{6} \qquad \text{(cancel numerator and denominator by } a\text{)}$$

$$\frac{3x^2}{15xy} = \frac{x}{5y} \qquad \text{(cancel numerator and denominator by } 3x\text{)}.$$

Exercise 48

Copy and complete the following.

1. $\dfrac{3}{4} = \dfrac{9}{-}$

2. $\dfrac{1}{5} = \dfrac{}{20}$

3. $\dfrac{2x}{3} = \dfrac{10x}{}$

4. $\dfrac{3}{8ab} = \dfrac{6}{-}$

5. $\dfrac{a}{5b} = \dfrac{ab}{}$

6. $\dfrac{xy}{3a} = \dfrac{5xyz}{}$

7. $\dfrac{t}{3s} = \dfrac{}{9st}$

8. $\dfrac{5a}{12b} = \dfrac{}{36ab^2}$

9. $\dfrac{15}{20} = \dfrac{3}{-}$

10. $\dfrac{8}{24} = \dfrac{}{3}$

11. $\dfrac{16y}{18} = \dfrac{8y}{}$

12. $\dfrac{5a}{30b} = \dfrac{}{6b}$

13. $\dfrac{xy}{ay} = \dfrac{x}{-}$

14. $\dfrac{3p}{7p} = \dfrac{}{7}$

15. $\dfrac{x^2y}{2xy} = \dfrac{}{2}$

16. $\dfrac{12ab}{3a} = \dfrac{4b}{-}$

17. $\dfrac{7x^2}{14x} = \dfrac{x}{-}$

18. $\dfrac{4abc}{9a^2b^2c^2} = \dfrac{4}{-}$

19. $\dfrac{30x}{12xy} = \dfrac{}{2y}$

20. $\dfrac{2a}{t} = \dfrac{}{t^2}$

21. $\dfrac{mn}{5p} = \dfrac{3m^2n}{}$

22. $\dfrac{x^2}{y^2} = \dfrac{}{x^2y^2}$

23. $\dfrac{1}{4x} = \dfrac{4y}{-}$

24. $\dfrac{4ab}{12} = \dfrac{}{3}$

Exercise 49

Reduce, by cancelling, each of the following fractions to their lowest terms.

1. $\dfrac{3a}{6b}$

2. $\dfrac{ax}{ay}$

3. $\dfrac{2t}{3t}$

4. $\dfrac{x^2}{xy}$

5. $\dfrac{2xy}{10ty}$

6. $\dfrac{7h}{21h^2}$

7. $\dfrac{5a}{10a}$

8. $\dfrac{xy}{xy}$

9. $\dfrac{2x^2yz}{3xy^2z}$ 11. $\dfrac{mn}{3m}$ 13. $\dfrac{5}{15x^2}$ 15. $\dfrac{x^2y^2}{xy}$

10. $\dfrac{ab}{3abc}$ 12. $\dfrac{xy}{5xy}$ 14. $\dfrac{22yz}{55xy}$

(b) Addition/subtraction

(i) In the most simple cases, where the denominators of the fractions are the same, the numerators may be added or subtracted directly.

Consider:

$$\frac{5}{12}+\frac{3}{12}=\frac{5+3}{12}=\frac{8}{12} \qquad \frac{5}{12}-\frac{3}{12}=\frac{5-3}{12}=\frac{2}{12}$$

$$\frac{4a}{3}+\frac{a}{3}=\frac{4a+a}{3}=\frac{5a}{3} \qquad \frac{4a}{3}-\frac{a}{3}=\frac{4a-a}{3}=\frac{3a}{3}$$

$$\frac{2x}{7}+\frac{3y}{7}=\frac{2x+3y}{7} \qquad \frac{2x}{7}-\frac{3y}{7}=\frac{2x-3y}{7}$$

$$\frac{3}{a}+\frac{2}{a}=\frac{3+2}{a}=\frac{5}{a} \qquad \frac{3}{a}-\frac{2}{a}=\frac{3-2}{a}=\frac{1}{a}$$

$$\frac{3h}{4x}+\frac{5m}{4x}=\frac{3h+5m}{4x} \qquad \frac{3h}{4x}-\frac{5m}{4x}=\frac{3h-5m}{4x}$$

Exercise 50

Simplify

1. $\dfrac{3}{4}+\dfrac{7}{4}$ 6. $\dfrac{9}{x}-\dfrac{3}{x}$ 11. $\dfrac{2x}{8}-\dfrac{2a}{8}$ 16. $\dfrac{x}{a^2}-\dfrac{y}{a^2}$

2. $\dfrac{2x}{3}+\dfrac{5x}{3}$ 7. $\dfrac{b}{x}-\dfrac{c}{x}$ 12. $\dfrac{3x}{ab}+\dfrac{5y}{ab}$ 17. $\dfrac{3}{abc}+\dfrac{6}{abc}$

3. $\dfrac{3}{a}+\dfrac{4}{a}$ 8. $\dfrac{2}{xy}+\dfrac{7}{xy}$ 13. $\dfrac{4x}{3}+\dfrac{4x}{3}$ 18. $\dfrac{2m}{4t}-\dfrac{3n}{4t}$

4. $\dfrac{7}{2}-\dfrac{3}{2}$ 9. $\dfrac{a}{bc}+\dfrac{5a}{bc}$ 14. $\dfrac{m}{2}-\dfrac{m}{2}$ 19. $\dfrac{7}{8x}-\dfrac{3}{8x}+\dfrac{5}{8x}$

5. $\dfrac{5x}{7}-\dfrac{2x}{7}$ 10. $\dfrac{a}{bc}+\dfrac{b}{bc}$ 15. $\dfrac{3ab}{x}+\dfrac{4ab}{x}$ 20. $\dfrac{2a}{x}+\dfrac{3b}{x}-\dfrac{a}{x}$

(ii) When the fractions to be added or subtracted have different denominators, then the fractions have to be altered so that the denominators become the same. This means finding a common multiple, and preferably the lowest common multiple (or LCM) for the denominators.

Consider:

$\frac{2}{3} + \frac{3}{4}$ LCM of 3 and 4 is 12, therefore

$$\frac{2}{3} + \frac{3}{4}$$
$$\downarrow \quad \downarrow$$
$$\frac{8}{12} + \frac{9}{12} = \frac{17}{12}.$$

$\frac{3x}{5} - \frac{y}{6}$ LCM of 5 and 6 is 30, therefore

$$\frac{3x}{5} - \frac{y}{6}$$
$$\downarrow \quad \downarrow$$
$$\frac{18x}{30} - \frac{5y}{30} = \frac{18x - 5y}{30}.$$

$\frac{4}{a} + \frac{3}{b}$ LCM of a and b is ab, therefore

$$\frac{4}{a} + \frac{3}{b}$$
$$\downarrow \quad \downarrow$$
$$\frac{4b}{ab} + \frac{3a}{ab} = \frac{4b + 3a}{ab}.$$

$\frac{2x}{pq} - \frac{5y}{pr}$ LCM of pq and pr is pqr, therefore

$$\frac{2x}{pq} - \frac{5y}{pr}$$
$$\downarrow \quad \downarrow$$
$$\frac{2rx}{pqr} - \frac{5qy}{pqr} = \frac{2rx - 5qy}{pqr}.$$

$\dfrac{2x}{a^2} + \dfrac{y}{3ab}$ LCM of a^2 and $3ab$ is $3a^2b$, therefore

$$\dfrac{2x}{a^2} + \dfrac{y}{3ab}$$

$$\downarrow \qquad \downarrow$$

$$\dfrac{6bx}{3a^2b} + \dfrac{ay}{3a^2b} = \dfrac{6bx + ay}{3a^2b}.$$

Exercise 51

Simplify.

1. $\dfrac{x}{2} - \dfrac{x}{3}$

2. $\dfrac{5x}{6} + \dfrac{2x}{3}$

3. $\dfrac{x}{4} - \dfrac{x}{5}$

4. $\dfrac{2x}{3} + \dfrac{x}{2} + \dfrac{3x}{4}$

5. $\dfrac{5x}{8} - \dfrac{2x}{5}$

6. $\dfrac{xy}{3} + \dfrac{3xy}{5}$

7. $\dfrac{x}{6} + \dfrac{3x}{4} - \dfrac{5x}{8}$

8. $\dfrac{4x}{21} + \dfrac{4x}{7}$

9. $\dfrac{7x}{9} - \dfrac{x}{2}$

10. $\dfrac{3}{x} + \dfrac{4}{y}$

11. $\dfrac{2x}{a} - \dfrac{y}{b}$

12. $\dfrac{a}{b} + \dfrac{b}{a}$

13. $\dfrac{3}{x^2} + \dfrac{4}{x}$

14. $\dfrac{5}{x} + \dfrac{3}{4}$

15. $\dfrac{6}{xy} - \dfrac{2}{yz}$

16. $\dfrac{5}{a} - 2$

17. $\dfrac{7a}{b} - \dfrac{7b}{a}$

18. $x + \dfrac{2y}{5}$

(iii) Consider:

Example 1

Simplify

$$\dfrac{3x - 2}{3} + \dfrac{2x + 5}{3}.$$

Since the numerators in each case represent a single quantity, they should be enclosed within brackets.
So we may write

$$\dfrac{(3x - 2)}{3} + \dfrac{(2x + 5)}{3}.$$

Now we can proceed as in the earlier examples, but taking care to observe the usual procedures for handling brackets.
Therefore

$$\frac{(3x-2)}{3} + \frac{(2x+5)}{3} = \frac{(3x-2)+(2x+5)}{3}$$

$$= \frac{3x-2+2x+5}{3}$$

$$= \frac{5x+3}{3}.$$

Example 2

$$\frac{(7x+3)}{4} - \frac{(3x-6)}{4} = \frac{(7x+3)-(3x-6)}{4}$$

$$= \frac{7x+3-3x+6}{4}$$

$$= \frac{4x+9}{4}.$$

Example 3

$$\frac{(2x+1)}{3} + \frac{(x-2)}{4} = \frac{4(2x+1)}{12} + \frac{3(x-2)}{12}$$

(LCM of 3 and 4 is 12)

$$= \frac{4(2x+1)+3(x-2)}{12}$$

$$= \frac{8x+4+3x-6}{12}$$

$$= \frac{11x-2}{12}.$$

Example 4

$$\frac{(x-2)}{5} - \frac{(2x+1)}{3} = \frac{3(x-2)}{15} - \frac{5(2x+1)}{15}$$

(LCM of 5 and 3 is 15)

$$= \frac{3(x-2)-5(2x+1)}{15}$$

$$= \frac{3x-6-10x-5}{15}$$

$$= \frac{-7x-11}{15}.$$

31

Exercise 52

Simplify.

1. $\dfrac{2x+1}{3} + \dfrac{x+4}{3}$

2. $\dfrac{5x-3}{4} + \dfrac{x+6}{4}$

3. $\dfrac{3x+2}{5} - \dfrac{2x+1}{5}$

4. $\dfrac{10x+5}{2} - \dfrac{3x-6}{2}$

5. $\dfrac{x-7}{10} + \dfrac{2x-9}{10}$

6. $\dfrac{3x+2}{6} - \dfrac{5x-2}{6}$

7. $\dfrac{x+2}{2} + \dfrac{2x+1}{3}$

8. $\dfrac{6x-1}{4} + \dfrac{3x-2}{3}$

9. $\dfrac{x+1}{2} - \dfrac{x+1}{4}$

10. $\dfrac{2x+3}{3} - \dfrac{x-5}{5}$

11. $\dfrac{2x-3}{3} + \dfrac{2x-5}{4}$

12. $\dfrac{3x}{4} - \dfrac{x+1}{5}$

13. $\dfrac{2x+3}{4} + 5x$

14. $\dfrac{5x-2}{6} - \dfrac{4x+2}{9}$

15. $\dfrac{3-x}{2} + \dfrac{2x+4}{3}$

16. $\dfrac{2x-1}{3} + \dfrac{4x+1}{4} - \dfrac{3x-2}{2}$

17. $\dfrac{5x+4}{3} + \dfrac{3}{4}$

18. $x - \dfrac{3x-6}{5}$

Harder examples

Consider:

1. $\dfrac{2(x-3)}{3} + \dfrac{3(x+4)}{4} = \dfrac{4 \times 2(x-3)}{12} + \dfrac{3 \times 3(x+4)}{12}$

$$\text{(LCM of 3 and 4 is 12)}$$

$$= \dfrac{8(x-3)}{12} + \dfrac{9(x+4)}{12}$$

$$= \dfrac{8(x-3)+9(x+4)}{12}$$

$$= \frac{8x - 24 + 9x + 36}{12}$$

$$= \frac{17x + 12}{12}.$$

2. $\dfrac{3}{x+2} - \dfrac{2}{x-3} \qquad = \dfrac{3(x-3)}{(x+2)(x-3)} - \dfrac{2(x+2)}{(x+2)(x-3)}$

(LCM of $x+2$ and $x-3$
is $(x+2)(x-3)$)

$$= \frac{3(x-3) - 2(x+2)}{(x+2)(x-3)}$$

$$= \frac{3x - 9 - 2x - 4}{(x+2)(x-3)}$$

$$= \frac{x-13}{(x+2)(x-3)}.$$

$\left(\text{Note: this answer may also be written as } \dfrac{x-13}{x^2-x-6}. \right)$

Exercise 53

Simplify.

1. $\dfrac{2(x+2)}{3} + \dfrac{(x-1)}{2}$

2. $\dfrac{3(2x-5)}{4} - \dfrac{2(x+1)}{3}$

3. $\dfrac{4(x+1)}{5} + \dfrac{3(2x-3)}{4}$

4. $\dfrac{2x}{7} - \dfrac{3(x-2)}{21}$

5. $\dfrac{5(2x+3)}{6} - \dfrac{4(x+2)}{5}$

6. $\dfrac{3(x-2)}{2} + \dfrac{3(2x+1)}{5}$

7. $\dfrac{1}{x+1} + \dfrac{1}{x-1}$

8. $\dfrac{3}{2x-3} + \dfrac{2}{x+3}$

9. $\dfrac{1}{2x-5} - \dfrac{1}{2x+5}$

10. $\dfrac{4}{x-5} + \dfrac{2}{2x+1}$

11. $\dfrac{x}{x-y} - \dfrac{y}{x+2y}$

12. $\dfrac{5}{2x} - \dfrac{6}{x+y}$

(c) Multiplication/division

Consider:

$$\frac{2}{3} \times \frac{1}{2} = \frac{2 \times 1}{3 \times 2} = \frac{2}{6} = \frac{1}{3} \qquad \text{or} \qquad \frac{1\cancel{2}}{3} \times \frac{1}{\cancel{2}_1} = \frac{1 \times 1}{3 \times 1} = \frac{1}{3}.$$

$$\frac{3}{4} \times \frac{2}{3} = \frac{3 \times 2}{4 \times 3} = \frac{6}{12} = \frac{1}{2} \qquad \text{or} \qquad \frac{1\cancel{3}}{2\cancel{4}} \times \frac{\cancel{2}^1}{\cancel{3}_1} = \frac{1 \times 1}{2 \times 1} = \frac{1}{2}.$$

$$\frac{a}{b^2} \times \frac{b}{c} = \frac{a \times b}{b^2 \times c} = \frac{ab}{b^2 c} = \frac{a}{bc} \qquad \text{or} \qquad \frac{a}{b\cancel{b}^2} \times \frac{\cancel{b}^1}{c} = \frac{a \times 1}{b \times c} = \frac{a}{bc}.$$

$$\frac{x^6}{y^2} \times \frac{y^3}{x^2} = \frac{x^6 \times y^3}{y^2 \times x^2} = \frac{x^6 y^3}{x^2 y^2} \qquad \text{or} \qquad \frac{x^4\cancel{x}^6}{1\cancel{y}^2} \times \frac{\cancel{y}^{3y}}{\cancel{x}^2{}_1} = \frac{x^4 \times y}{1 \times 1} = \frac{x^4 y}{1}.$$

$$= \frac{x^4 y}{1}$$

Exercise 54

Simplify.

1. $\dfrac{x}{2} \times \dfrac{4}{y}$

2. $\dfrac{xy}{a} \times \dfrac{a}{5}$

3. $\dfrac{3x}{5} \times \dfrac{10}{9y}$

4. $\dfrac{xy}{a} \times \dfrac{a^2}{2x}$

5. $\dfrac{x}{2y} \times \dfrac{y}{2x}$

6. $\dfrac{a^2}{x^2} \times \dfrac{x}{ab}$

7. $5x \times \dfrac{3y}{20}$

8. $\dfrac{a}{b} \times \dfrac{b}{c} \times \dfrac{c}{d}$

9. $\dfrac{2x^2}{3} \times \dfrac{y}{x^2}$

10. $\dfrac{a}{2b} \times 4$

11. $\dfrac{7x^2}{8yz} \times \dfrac{2y^2}{21xz}$

12. $\dfrac{ab}{c} \times \dfrac{a^2}{b} \times \dfrac{3}{a}$

13. $\dfrac{ab}{x^2 y^2} \times \dfrac{xy}{a^2 b^2}$

14. $\dfrac{2x}{5} \times \dfrac{10}{3y} \times \dfrac{6x}{22y}$

15. $\dfrac{4x^2}{3} \times \dfrac{2}{y} \times \dfrac{xy}{2x^2}$

16. $\dfrac{3a}{2} \times ab \times \dfrac{2}{ab^2}$

Consider:

$$\frac{2}{3} \div \frac{2}{5} \quad \text{becomes} \quad \frac{2}{3} \times \frac{5}{2} \quad \text{i.e.} \quad \frac{1\cancel{2}}{3} \times \frac{5}{\cancel{2}_1} = \frac{5}{3}.$$

$$\frac{a^2}{b} \div \frac{2a}{3b} \quad \text{becomes} \quad \frac{a^2}{b} \times \frac{3b}{2a} \quad \text{i.e.} \quad \frac{{}^{a}\cancel{a^2}}{{}_{1}\cancel{b}} \times \frac{3\cancel{b}}{2\cancel{a}} = \frac{3a}{2}.$$

$$\frac{xy}{2} \div \frac{x^2}{6y} \times \frac{2x}{5y} \quad \text{becomes} \quad \frac{xy}{2} \times \frac{6y}{x^2} \times \frac{2x}{5y} \quad \text{i.e.} \quad \frac{{}^{1}\cancel{x}y}{{}_{1}\cancel{2}} \times \frac{6\cancel{y}}{\cancel{x^2}\,_{1}} \times \frac{\cancel{2}\cancel{x}^{1}}{5\cancel{y}}$$

$$= \frac{6y}{5}.$$

Exercise 55

Simplify.

1. $\dfrac{3}{x} \div \dfrac{6y}{x^2}$

2. $\dfrac{ab}{2} \div \dfrac{bc}{a}$

3. $\dfrac{5}{xy} \div \dfrac{10}{3x}$

4. $x \div \dfrac{xy}{z}$

5. $\dfrac{2x}{3y} \div \dfrac{6y}{5x} \times \dfrac{3x^2}{y^2}$

6. $\dfrac{5x}{6y} \div \dfrac{y}{2x} \div \dfrac{x^2}{y^2}$

7. $\dfrac{1}{7xy} \div \dfrac{2}{21x^2}$

8. $\dfrac{3x}{10y} \div 6x^2$

9. $\dfrac{x}{2y} \div \dfrac{4x^2}{3y^2}$

10. $\dfrac{5x}{2y} \div 10x \div \dfrac{1}{4y^2}$

2 Equations – solution by calculation

This chapter deals with the solution of simple, simultaneous (linear), and quadratic equations, by calculation, that is to say, the re-arrangement and manipulation of the terms contained in an equation, in order to find its solution.

2.1 Simple equations

Simple equations are expressions which contain one unknown quantity. SOLVING a simple equation is the process of finding the unique value of the unknown quantity which will satisfy the expression, that is, make the expression a true statement.

Consider the following basic situations:

1. $x + 3 = 7$
   $x + 3 - 3 = 7 - 3$ (*subtract* 3 from *both* sides)
         $x = 4$.
   Check: $4 + 3 = 7$ *true*.

2. $x - 1 = 5$
   $x - 1 + 1 = 5 + 1$ (*add* 1 to *both* sides)
         $x = 6$.
   Check: $6 - 1 = 5$ *true*.

3. $3x = 6$

   $$x = \frac{6}{3} \quad (\textit{divide both} \text{ sides by 3})$$

       $x = 2$.
   Check: $3 \times 2 = 6$ *true*.

4. $\dfrac{x}{2} = 7$

$x = 7 \times 2$ (*multiply both* sides by 2)

$x = 14.$

Check: $\dfrac{14}{2} = 7$ *true.*

(i) Whatever process – addition/subtraction/multiplication/division – is used, it must be applied to the *whole equation*.

(ii) The solution arrived at always makes the original statement a *true* statement. No other solution would be acceptable.

Exercise 56

Solve the following, and in each case *check* that the solution satisfies the original statement.

1. $x + 6 = 9$	11. $x - 5 = 3$	21. $5x = 15$
2. $x + 3 = 11$	12. $x - 2 = 12$	22. $3x = 30$
3. $x + 7 = 14$	13. $x - 1 = 4$	23. $7x = 28$
4. $x + 5 = 13$	14. $x - 10 = 10$	24. $4x = 16$
5. $2 + x = 3$	15. $x - 6 = 0$	25. $2x = 9$
6. $7 + x = 12$	16. $x - 3 = 2\frac{1}{2}$	26. $3x = 10$
7. $x + 3 = 5\frac{1}{2}$	17. $x - 6\frac{1}{2} = 20$	27. $6x = 48$
8. $6 + x = 7\frac{1}{2}$	18. $x - 3\frac{1}{2} = 4\frac{1}{2}$	28. $10x = 5$
9. $x + 12 = 12$	19. $x - 2 = -1$	29. $5x = 3$
10. $7 + x = 2$	20. $5 - x = 3$	30. $9x = 21$
31. $\dfrac{x}{4} = 5$	36. $\dfrac{x}{10} = 5$	
32. $\dfrac{x}{3} = 3$	37. $\dfrac{x}{5} = 1\frac{1}{2}$	
33. $\dfrac{x}{6} = 1$	38. $\dfrac{x}{3} = 16$	
34. $\dfrac{x}{7} = 3$	39. $\dfrac{x}{5} = 0$	
35. $\dfrac{x}{2} = 3\frac{1}{2}$	40. $\dfrac{x}{8} = 4$	

Consider:

1. $2x + 4 = 18$
   $2x = 18 - 4$
   $2x = 14$
   $x = \dfrac{14}{2}$
   $x = 7.$

2. $3x + 4 = 5x + 7$
   $3x - 5x = +7 - 4$
   $-2x = +3$
   $x = \dfrac{+3}{-2}$
   $x = -1\frac{1}{2}.$

3. $6x - 1 = 3 - 4x$
   $6x + 4x = 3 + 1$
   $10x = 4$
   $x = \dfrac{4}{10}$
   $x = \frac{2}{5}.$

Exercise 57

Solve the following equations.

1. $3x - 7 = 11$
2. $2x + 19 = 30$
3. $3x + 2 = x + 8$
4. $15x - 3 = 6x + 12$
5. $4x + 5 = 3x - 6$
6. $x - 6 = 3 - 2x$
7. $x - 10 = 3x - 6$
8. $20 - 7x = 3x + 10$
9. $5x = 12 + 3x$
10. $6 + 2x = x + 6$
11. $12 = 20 - 4x$
12. $7x + 3 = 16 - 6x$
13. $5 - 4x = 23 - 7x$
14. $6x + 4 = 8x - 10$
15. $13x - 7 = 8x + 10$
16. $7 + 8x = 4x + 31$
17. $1 + 6x - 10 = x - 9$
18. $8x = 11x + 16 - 8x$
19. $4x - 6 + x = 3x + 14$
20. $6 - 5x = 8x - 7$
21. $9x - 32 = 5x - 13 - 6x + 11$
22. $13x + 7 - 8x = 32$
23. $x = 6 + 2x + 3$
24. $11 - 4x = 7 - 7x + 2$

Consider:

1. $3(x - 4) + 2(2x + 1) = 4$
   $3x - 12 + 4x + 2 = 4$
   $7x - 10 = 4$
   $7x = 4 + 10$
   $7x = 14$
   $x = \dfrac{14}{7}$
   $x = 2.$

2. $14 - x = 2(x + 2) - 5(x - 6)$
   $14 - x = 2x + 4 - 5x + 30$
   $14 - x = -3x + 34$
   $3x - x = 34 - 14$
   $2x = 20$
   $x = \dfrac{20}{2}$
   $x = 10.$

Exercise 58

Solve.

1. $2(x - 2) + 2(3x + 1) = 6$
2. $5(1 - 2x) - 3(x + 1) = 28$
3. $3 = 6(x + 2) - (x - 1)$
4. $3(x - 2) + 5(2x + 3) = 11x + 15$
5. $4(5x - 2) + 3(5 - 6x) = 0$
6. $6(1 + x) - 2(2x - 3) = 20$
7. $7(x - 6) = 3(2x - 4)$
8. $2(3x + 2) - 3(2x + 1) = x$
9. $(2x + 1) - (x - 2) + (x - 5) = 0$
10. $5(4x + 3) - 3(6x + 10) = 1$
11. $2(2x - 1) + 2(4 - 3x) = 0$
12. $x + 3(2x - 1) = 4(x + 3)$

13. $2(2x-5)-3(x-6)=10$ 15. $2(x-1)-3(x+4)=3(x+2)$
14. $4(1-2x)+6(x+3)=30$ 16. $0=6(2x-5)-3(3x-7)$

Consider:

1. $\dfrac{x}{5}=10$ 2. $\dfrac{2x}{5}=10$ 3. $\dfrac{3x}{4}=8$

$x=10\times 5$ $2x=10\times 5$ $3x=8\times 4$

$x=50.$ $x=\dfrac{10\times 5}{2}$ $x=\dfrac{8\times 4}{3}$

$x=25.$ $x=\dfrac{32}{3}$

$=10\tfrac{2}{3}.$

Exercise 59

Solve.

1. $\dfrac{x}{6}=2$ 4. $\dfrac{2x}{5}=2$ 7. $\dfrac{-2x}{3}=14$ 10. $\dfrac{2x}{5}=5$

2. $\dfrac{2x}{3}=6$ 5. $\dfrac{5x}{6}=10$ 8. $\dfrac{3x}{5}=9$ 11. $\dfrac{-3x}{4}=-3$

3. $\dfrac{3x}{4}=9$ 6. $\dfrac{x}{4}=1\tfrac{1}{2}$ 9. $\dfrac{7x}{10}=14$ 12. $\dfrac{4x}{3}=12$

Consider:

1. $\dfrac{x}{3}+\dfrac{x}{2}=10$ 2. $\dfrac{2x}{5}-\dfrac{x}{2}=1$ 3. $\dfrac{3x}{4}-\dfrac{x}{2}=\dfrac{4}{3}$

LCM of 3 and 2 is 6.   LCM of 5 and 2 is 10.   LCM of 4, 2 and 3 is 12.

Multiply equation by 6   *Multiply* equation by 10   *Multiply* equation by 12

$\dfrac{6\times x}{3}+\dfrac{6\times x}{2}$ $\dfrac{10\times 2x}{5}-\dfrac{10\times x}{2}$ $\dfrac{12\times 3x}{4}-\dfrac{12\times x}{2}$

$=6\times 10$ $=10\times 1$ $=\dfrac{12\times 4}{3}$

$2x+3x=60$ $4x-5x=10$ $9x-6x=16$

$5x=60$ $-x=10$ $3x=16$

$x=\dfrac{60}{5}=12.$ $x=-10$ $x=\dfrac{16}{3}$

$=5\tfrac{1}{3}.$

39

Exercise 60

Solve.

1. $\dfrac{x}{2} + \dfrac{x}{3} = 5$
4. $\dfrac{3x}{4} - \dfrac{x}{3} = 5$
7. $\dfrac{x}{5} + \dfrac{x}{6} - \dfrac{x}{2} = 2$

2. $\dfrac{2x}{3} + \dfrac{x}{2} = 1$
5. $\dfrac{x}{2} - \dfrac{x}{4} = 2$
8. $\dfrac{x}{2} - \dfrac{3x}{5} = -1$

3. $\dfrac{x}{4} + \dfrac{x}{5} = 9$
6. $\dfrac{3x}{2} - \dfrac{2x}{3} = 2\tfrac{1}{2}$
9. $\dfrac{2x}{5} - \dfrac{3x}{4} + \dfrac{x}{2} = 1$

Consider:

1. $\dfrac{x+2}{2} + \dfrac{2x-1}{3} = 3$

LCM of 2 and 3 is 6.
Multiply equation by 6

$$\dfrac{6(x+2)}{2} + \dfrac{6(2x-1)}{3} = 6 \times 3$$

$$3(x+2) + 2(2x-1) = 18$$
$$3x + 6 + 4x - 2 = 18$$
$$7x + 4 = 18$$
$$7x = 18 - 4$$
$$7x = 14$$
$$x = \dfrac{14}{7} = 2.$$

2. $\dfrac{4x+1}{5} - \dfrac{x-2}{2} = \dfrac{9}{10}$

LCM of 5, 2 and 10 is 10.
Multiply equation by 10

$$\dfrac{10(4x+1)}{5} - \dfrac{10(x-2)}{2} = 10 \times \dfrac{9}{10}$$

$$2(4x+1) - 5(x-2) = 9$$
$$8x + 2 - 5x + 10 = 9$$
$$3x + 12 = 9$$
$$3x = 9 - 12$$
$$3x = -3$$
$$x = \dfrac{-3}{3} = -1.$$

Exercise 61

Solve.

1. $\dfrac{x+1}{2} + \dfrac{x+2}{3} = 2$
5. $\dfrac{x+4}{2} - \dfrac{x-2}{4} = 3$

2. $\dfrac{x-3}{5} - \dfrac{x-2}{2} = 1$
6. $\dfrac{2x-1}{6} - \dfrac{x+2}{5} = \dfrac{11}{10}$

3. $\dfrac{x-2}{6} + \dfrac{x-1}{4} = \dfrac{2}{3}$
7. $\dfrac{x-2}{3} - \dfrac{x+3}{5} = \dfrac{1}{5}$

4. $\dfrac{x+5}{4} - \dfrac{x+4}{5} = 0$
8. $\dfrac{x-4}{3} = \dfrac{x+3}{4}$

9. $\dfrac{3x+2}{2}+\dfrac{2x-1}{3}=5$

11. $\dfrac{x+1}{6}+\dfrac{x+4}{3}=0$

10. $\dfrac{x+3}{2}-\dfrac{5-x}{5}=-\dfrac{1}{5}$

12. $\dfrac{1-2x}{5}+\dfrac{2x-5}{4}=\dfrac{3}{20}$

Consider:

1. $\dfrac{3}{x+1}=\dfrac{4}{x-2}$

LCM of $x+1$ and $x-2$
is $(x+1)(x-2)$.
Multiply equation
by $(x+1)(x-2)$

$(x+1)(x-2)\dfrac{3}{x+1}=$

$(x+1)(x-2)\dfrac{4}{x-2}$

$3(x-2)=4(x+1)$
$3x-6=4x+4$
$3x-4x=4+6$
$-x=10$
$x=-10.$

2. $\dfrac{4(2x-1)}{5}-\dfrac{3(x+6)}{4}=-7$

LCM of 5 and 4 is 20.

Multiply equation by 20

$\dfrac{20\times 4(2x-1)}{5}-\dfrac{20\times 3(x+6)}{4}=20\times -7$

$16(2x-1)-15(x+6)=-140$
$32x-16-15x-90=-140$
$17x-106=-140$
$17x=-140+106$
$17x=-34$
$x=-\dfrac{34}{17}=-2.$

Exercise 62

Solve.

1. $\dfrac{3}{x-2}=\dfrac{2}{x-3}$

6. $\dfrac{1}{2x+1}-\dfrac{5}{x-3}=0$

2. $\dfrac{4}{x-1}=\dfrac{3}{x+1}$

7. $\dfrac{3(x+6)}{4}-\dfrac{2(2x+3)}{3}=-1$

3. $\dfrac{5}{3x+2}=\dfrac{3}{2x+3}$

8. $\dfrac{4(x+1)}{3}-\dfrac{(3x-5)}{2}=3$

4. $\dfrac{3}{5x-4}=\dfrac{7}{11x+4}$

9. $\dfrac{3(3x+8)}{2}=3+\dfrac{3(4x+1)}{5}$

5. $\dfrac{2}{5-x}=\dfrac{3}{x+2}$

10. $\dfrac{2(3x+1)}{5}+\dfrac{3(3x-1)}{4}=10$

2.2   Simultaneous equations (linear)

Simultaneous equations, each with two unknown quantities, are handled in pairs. In solving such equations, we are trying to find a value for each unknown quantity that will satisfy *both* equations *at the same time*, that is simultaneously!

Consider the two equations $x + y = 5$ and $x - y = 1$. Can we find a value for x and a value for y to satisfy both equations?

$x + y = 5$	$x - y = 1$
$x = 4, \ y = 1$	$x = 5, \ y = 4$
$x = 3, \ y = 2$	$x = 4, \ y = 3$
$x = 2, \ y = 3$	$x = 3, \ y = 2$
$x = 1, \ y = 4$	$x = 2, \ y = 1$

These are just some of the values of x and y which satisfy the equations separately. These lists could be extended indefinitely. However, *one pair* of values occurs in both lists, $x = 3$, $y = 2$. These values satisfy both equations at the same time (simultaneously).

No other values will do this. So we may say that:

equations $\quad \left. \begin{matrix} x + y = 5 \\ x - y = 1 \end{matrix} \right\} \quad$ are satisfied by $\quad \left. \begin{matrix} x = 3 \\ y = 2 \end{matrix} \right\}$

The above method of finding the values (guesswork) will not be suitable for more complicated examples. We shall now consider a proper procedure.

Elimination

If we can combine the two equations in such a way as to eliminate one variable, then we shall be left with a simple equation – from which we can find the other variable. Having found one of the variables it should be a straightforward matter to find the other.

There are a variety of methods of achieving the elimination necessary – but for simplicity the method used here will always be that of *adding together the two equations.*

Consider:

Example 1

$$2x + y = 11 \quad \ldots \ldots \text{ equation (1)}$$
$$x - y = 1 \quad \ldots \ldots \text{ equation (2)}$$

Add equation (1) and equation (2)

$$3x = 12$$
$$x = \frac{12}{3} = 4.$$

Substitute $x = 4$ in equation (2)

$$(4) - y = 1$$
$$4 - y = 1$$
$$-y = 1 - 4$$
$$-y = -3$$
$$y = 3.$$

Answer: $\left. \begin{array}{l} x = 4 \\ y = 3 \end{array} \right\}$

Note (i) $x = 4$ could have been substituted in equation (1). Try this, you will still get $y = 3$.
(ii) Check that the solutions satisfy both equations.

Example 2

$$2x + y = 1 \quad \ldots \ldots \quad (1)$$
$$3x - y = 9 \quad \ldots \ldots \quad (2)$$

Add (1) and (2)

$$5x = 10$$
$$x = \frac{10}{5} = 2.$$

Substitute $x = 2$ in equation (1)

$$2(2) + y = 1$$
$$4 + y = 1$$
$$y = 1 - 4$$
$$y = -3.$$

Answer: $\left. \begin{array}{l} x = 2 \\ y = -3 \end{array} \right\}$

Example 3

$$-x + 4y = 3 \quad \ldots \ldots \quad (1)$$
$$x - 3y = -1 \quad \ldots \ldots \quad (2)$$

Add (1) and (2)

$$y = 2.$$

Substitute $y = 2$ in equation (2)

$$x - 3(2) = -1$$
$$x - 6 = -1$$
$$x = -1 + 6$$
$$x = 5.$$

Answer: $\left.\begin{array}{l} x = 5 \\ y = 2 \end{array}\right\}$

Exercise 63

Solve the following pairs of equations.

1. $x + y = 7$
   $x - y = 5$
2. $3x - y = 4$
   $2x + y = 11$
3. $x + 3y = -2$
   $-x - y = -2$

4. $4x + y = 1$
   $2x - y = 2$
5. $-x + y = 11$
   $5x - y = -23$
6. $x + y = -7$
   $3x - y = -5$

7. $2x + 3y = 7$
   $5x - 3y = 7$
8. $x - 2y = -3$
   $3x + 2y = 31$
9. $-2x + y = -14$
   $2x - 3y = 34$

Note You will have noticed that in every example so far the required elimination is possible because the equations contain terms which are numerically the same but of opposite sign; for example $+y$ and $-y$ *or* $-2x$ and $+2x$ *or* $+3y$ and $-3y$. Such terms are immediately eliminated by addition. However, it is not always possible to add directly.

Consider:

Example 1

$$x + y = 8 \qquad \ldots \ldots \quad (1)$$
$$2x + y = 11 \qquad \ldots \ldots \quad (2)$$

Add (1) and (2)

$$3x + 2y = 19.$$

This equation has no unique solution since it contains two variables! *Start again*

$$x + y = 8 \qquad \ldots \ldots \quad (1)$$
$$2x + y = 11 \qquad \ldots \ldots \quad (2)$$

Multiply equation (1) by -1.
The effect will be to *reverse all the signs*.

Hence
$$-x - y = -8 \qquad \ldots \ldots \quad (3)$$
$$2x + y = 11 \qquad \ldots \ldots \quad (2)$$

Add (3) and (2)

$$x = 3.$$

Substitute $x = 3$ in equation (2)

$$2(3) + y = 11$$
$$6 + y = 11$$
$$y = 11 - 6$$
$$y = 5.$$

Answer: $\left. \begin{array}{l} x = 3 \\ y = 5 \end{array} \right\}$

Note (i) Multiplying equation (2) by -1 would have produced the same results.

(ii) Check that the solutions satisfy equations (1) and (2).

Example 2

$$2x + y = 1 \qquad \ldots \ldots \quad (1)$$
$$2x - 3y = -11 \qquad \ldots \ldots \quad (2)$$

Multiply equation (2) by -1

Hence

$$-2x + 3y = 11 \qquad \ldots \ldots \quad (3)$$
$$2x + y = 1 \qquad \ldots \ldots \quad (1)$$

Add (3) and (1)

$$4y = 12$$
$$y = \frac{12}{4} = 3.$$

Substitute $y = 3$ in equation (1)

$$2x + (3) = 1$$
$$2x + 3 = 1$$
$$2x = 1 - 3$$
$$2x = -2$$
$$x = \frac{-2}{2} = -1.$$

Answer: $\left. \begin{array}{l} x = -1 \\ y = 3 \end{array} \right\}$

Exercise 64

Solve the following pairs of equations. You will need to multiply *one* of the equations by -1 in each case.

1. $2x + y = 5$
   $x + y = 3$
2. $3x - y = 7$
   $2x - y = 3$
3. $x + 2y = -3$
   $x - y = 6$

4. $4x + y = 4$
   $2x + y = 2\frac{1}{2}$
5. $5x + 6y = -1$
   $5x - y = -29$
6. $x + y = 17$
   $3x + y = 31$

7. $3x - y = -1$
   $3x + 2y = -16$
8. $5x + 4y = 2$
   $5x + 12y = 4$
9. $2x - 7y = 15$
   $3x - 7y = 19$

Harder examples

It is sometimes necessary to multiply one of the equations by a number other than -1 in order to create terms which are numerically alike but of opposite sign.

Consider:

Example 1

$$3x + 2y = 4 \quad \ldots\ldots \quad (1)$$
$$x - y = 3 \quad \ldots\ldots \quad (2)$$

There are two alternatives:
   (i) multiply equation (2) by 2 (to eliminate y)
or (ii) multiply equation (2) by -3 (to eliminate x).

Using alternative (i)

$$3x + 2y = 4 \quad \ldots\ldots \quad (1)$$
$$x - y = 3 \quad \ldots\ldots \quad (2)$$

Multiply equation (2) by 2

$$2x - 2y = 6 \quad \ldots\ldots \quad (3)$$
$$3x + 2y = 4 \quad \ldots\ldots \quad (1)$$

Add (3) and (1)

$$5x = 10$$

$$x = \frac{10}{5} = 2.$$

Substitute $x = 2$ in equation (2)

$$(2) - y = 3$$
$$2 - y = 3$$
$$-y = 3 - 2$$
$$-y = 1$$
$$y = -1.$$

Answer: $x = 2$ $\left.\right\}$
$\quad\quad\quad y = -1$

Example 2

$$2x - 3y = -8 \quad \ldots \ldots \quad (1)$$
$$4x - y = 14 \quad \ldots \ldots \quad (2)$$

The two alternatives are:
   (i) multiply equation (1) by -2 (to eliminate x)
or (ii) multiply equation (2) by -3 (to eliminate y).

Using alternative (i)

$$-4x + 6y = 16 \quad \ldots \ldots \quad (3)$$
$$4x - y = 14 \quad \ldots \ldots \quad (2)$$

Add (3) and (2)

$$5y = 30$$

$$y = \frac{30}{5} = 6.$$

Substitute $y = 6$ in equation (1)

$$2x - 3(6) = -8$$
$$2x - 18 = -8$$
$$2x = -8 + 18$$
$$2x = 10$$
$$x = \frac{10}{2} = 5.$$

Answer: $x = 5$ $\left.\right\}$
$\quad\quad\quad y = 6$

Exercise 65

Solve the following pairs of equations.

1. $4x - y = 1$
   $x + 3y = 10$
2. $2x + y = 9$
   $5x + 2y = 20$
3. $3x + 2y = 14$
   $x + y = 5$

4. $x + 2y = -7$
   $-2x + y = 4$
5. $x + y = 4\frac{1}{2}$
   $2x - 3y = 1\frac{1}{2}$
6. $2x + y = 1$
   $x - 3y = -10$

7. $x + y = 8$
   $2x + 3y = 19$
8. $4x - y = 11$
   $x + 2y = -13$
9. $2x + y = -1$
   $x + 2y = 1$

Further situations

Other situations may involve (i) multiplying both equations by different numbers, or (ii) equations which are fractional, or (iii) equations which contain brackets, or (iv) equations which need re-arranging. Whatever the situation, it is only the first stage that needs extra consideration.

Once the position has been reached where adding the equations eliminates a variable, the same procedure is always used to complete the solution.

Consider:

Example 1

$$2x + 3y = 11 \quad \ldots \ldots \quad (1)$$
$$3x - 2y = -3 \quad \ldots \ldots \quad (2)$$

Multiply equation (1) by 2, and equation (2) by 3

$$4x + 6y = 22 \quad \ldots \ldots \quad (3)$$
$$9x - 6y = -9 \quad \ldots \ldots \quad (4)$$

We can now add (3) and (4) and proceed to a solution.

Answer: $\left. \begin{array}{l} x = 1 \\ y = 3 \end{array} \right\}$

Example 2

$$3x + 2y - 16 = 0 \quad \ldots \ldots \quad (1)$$
$$4x = 2y - 2. \ldots \ldots \quad (2)$$

Re-arrange both equations

$$3x + 2y = 16 \quad \ldots \ldots \quad (1)$$
$$4x - 2y = -2 \quad \ldots \ldots \quad (2)$$

48

We can now add the equations and proceed to a solution.

Answer: $\left.\begin{array}{l} x = 2 \\ y = 5 \end{array}\right\}$

Example 3

$$2(x-4)+3y = 0 \quad \ldots \ldots \quad (1)$$
$$x-(5y-22) = 0 \quad \ldots \ldots \quad (2)$$

Expand the brackets

$$2x-8+3y = 0 \quad \ldots \ldots \quad (1)$$
$$x-5y+22 = 0 \quad \ldots \ldots \quad (2)$$

Re-arrange both equations

$$2x+3y = 8 \quad \ldots \ldots \quad (1)$$
$$x-5y = -22 \quad \ldots \ldots \quad (2)$$

Multiply equation (2) by -2

$$-2x+10y = 44 \quad \ldots \ldots \quad (3)$$
$$2x+3y = 8 \quad \ldots \ldots \quad (1)$$

We can now add and proceed to a solution.

Answer: $\left.\begin{array}{l} x = -2 \\ y = 4 \end{array}\right\}$

Example 4

$$\frac{x}{3}+\frac{y}{2} = 3 \quad \ldots \ldots \quad (1)$$

$$\frac{x}{2}-y = 1 \quad \ldots \ldots \quad (2)$$

To remove the fractions
   multiply equation (1) by 6 (LCM of 2 and 3)
   and equation (2) by 2 (LCM of 2 and 1)

$\left(\dfrac{6 \times x}{3}+\dfrac{6 \times y}{2} = 6 \times 3\right) \qquad 2x+3y = 18 \quad \ldots \ldots \quad (3)$

$\left(\dfrac{2 \times x}{2}-2 \times y = 2 \times 1\right) \qquad x-2y = 2 \quad \ldots \ldots \quad (4)$

Now multiply equation (4) by -2

$$-2x+4y = -4 \quad \ldots \ldots \quad (5)$$
$$2x+3y = 18 \quad \ldots \ldots \quad (3)$$

A—D

We can now add the equations and proceed to a solution.

Answer: $\left.\begin{array}{l} x = 6 \\ y = 2 \end{array}\right\}$

Exercise 66

Solve the following pairs of equations.

1. $2x = 7 + 3y$
   $x + y - 16 = 0$
2. $3x + 2y = -6$
   $2x + 3y = 1$
3. $\dfrac{x}{4} - \dfrac{y}{3} = 1$
   $\dfrac{x + 2y}{2} = 7$
4. $3(x - 1) = 6 - 2y$
   $4 - (x + y) = 0$
5. $4x + 3y = -13$
   $3x - 2y = 11\frac{1}{2}$
6. $2y = 8 - 5x$
   $10x + 5 = 3y$
7. $x - \dfrac{2y}{3} = 2$
   $\dfrac{4x}{5} - \dfrac{y}{2} = \dfrac{17}{10}$
8. $3x + 2y = -7$
   $3(2x - y) - 2(x + y) = 6$
9. $\dfrac{x + y}{4} = \frac{1}{2}$
   $\dfrac{x}{3} - y = 2$

2.3 Quadratic equations (by factorisation)

A quadratic equation is an equation that contains an x^2 term, but no higher powers of x. So that

$x^2 = 0, x^2 - 5x = 0, x^2 - 5x + 6 = 0, x^2 - 16 = 0, 3x^2 - 13x + 4 = 0$ are all quadratic equations.

Consider the quadratic equation $x^2 + 4x = 0$. This may be factorised and rewritten as

$x(x + 4) = 0.$

Now since $x(x + 4)$ is a *product* equal to zero, we may say that

either $x = 0$ *or* $(x + 4) = 0$
    i.e. $x + 4 = 0$
    i.e. $x = -4.$

So that the quadratic equation $x^2 + 4x = 0$ has two solutions (or roots), $x = 0$ or $x = -4$.

Example 2

Solve the quadratic equation $x^2 + x - 6 = 0.$
Factorising $(x + 3)(x - 2) = 0.$

Therefore, *either* $\qquad\qquad\qquad x+3=0$ *or* $x-2=0$
i.e. $\qquad\qquad\qquad\qquad\qquad x=-3$ or $x=2$.

Example 3

Solve $x^2 - 16 = 0$.
Note the 'difference of two squares'.

Factorising $\qquad\qquad\qquad (x+4)(x-4)=0$.
Therefore $\qquad\qquad\qquad x+4=0$ or $x-4=0$
i.e. $\qquad\qquad\qquad\qquad x=-4$ or $x=4$.

Example 4

Solve $x^2 + 6x + 9 = 0$
Note the perfect square.

Factorising $\qquad\qquad\qquad (x+3)(x+3)=0$.
Therefore $\qquad\qquad\qquad x+3=0$ or $x+3=0$
i.e. $\qquad\qquad\qquad\qquad x=-3$ or $x=-3$.

Although the same, the two solutions are treated as being separate but equal.

Note (i) The quadratic expression is *equated to zero*.
     (ii) It must factorise. (For this method to apply.)
    (iii) There are *two* solutions (or roots). They may be equal.
    (iv) The two solutions are separate.
     (v) Both solutions satisfy the equation.

Exercise 67

Solve the following quadratic equations, which are already in factor form.

1. $x(x-2)=0$
2. $x(x+3)=0$
3. $(x-1)(x+4)=0$
4. $(x+2)(x+2)=0$
5. $x(x-10)=0$
6. $(x+4)(x-5)=0$
7. $(x-6)(x-6)=0$
8. $x(x+5)=0$
9. $(x-1)(x+1)=0$
10. $(x+3)(x+6)=0$
11. $(3-x)(2+x)=0$
12. $(x+9)(x-7)=0$
13. $x(5-x)=0$
14. $(2+x)(x-7)=0$
15. $(x+3)^2=0$
16. $(x+4)(x-2)=0$
17. $(x-1)(x-10)=0$
18. $(x-7)^2=0$
19. $x(x+6)=0$
20. $(7-x)(7+x)=0$
21. $(x+8)(x+3)=0$
22. $(x+9)(x+11)=0$
23. $(1+x)(x-4)=0$
24. $(x+2)^2=0$

Exercise 68

Solve the following (you will need to factorise first).

1. $x^2 + 3x = 0$
2. $x^2 - 12x = 0$
3. $x^2 - 7x = 0$
4. $x^2 + x = 0$
5. $x^2 + 2x = 0$

6. $x^2 + 5x + 6 = 0$
7. $x^2 + 5x + 4 = 0$
8. $x^2 + 7x + 10 = 0$
9. $x^2 + 15x + 56 = 0$
10. $x^2 + 7x + 6 = 0$

11. $x^2 - 6x + 5 = 0$
12. $x^2 - 6x + 8 = 0$
13. $x^2 - 9x + 18 = 0$
14. $x^2 - 7x + 12 = 0$
15. $x^2 - 12x + 35 = 0$

16. $x^2 + 2x - 3 = 0$
17. $x^2 + x - 56 = 0$
18. $x^2 - 3x - 10 = 0$
19. $x^2 + 5x - 36 = 0$
20. $x^2 - x - 6 = 0$

21. $x^2 - 9 = 0$
22. $x^2 - 100 = 0$
23. $x^2 - 49 = 0$
24. $25 - x^2 = 0$
25. $x^2 - 169 = 0$

26. $x^2 + 6x + 9 = 0$
27. $x^2 - 2x + 1 = 0$
28. $x^2 - 10x + 25 = 0$
29. $x^2 + 20x + 100 = 0$
30. $x^2 - 16x + 64 = 0$

Example 1

Find the equation whose roots are 4 and -1.

We can say that	$x = 4$ or $x = -1$
i.e.	$x - 4 = 0$ or $x + 1 = 0$.
Therefore	$(x-4)(x+1) = 0$
i.e.	$x^2 - 3x - 4 = 0$.

Example 2

Find the equation whose roots are 0 and 2.

We can say that	$x = 0$ or $x = 2$
i.e.	$x = 0$ or $x - 2 = 0$.
Therefore	$x(x-2) = 0$
i.e.	$x^2 - 2x = 0$.

Exercise 69

Find the equations whose roots are:

1. 5, 2
2. 3, -1
3. -4, 2
4. -1, -5

5. 2, -2
6. -6, -6
7. 0, -5
8. 6, 0

9. 2, -7
10. 3, 4
11. -5, 6
12. 8, 0

13. -7, $+7$
14. 3, -5
15. -6, $+6$

Harder examples

1. Solve $3x^2 + 2x = 0$.
Factorising $x(3x + 2) = 0$.
Therefore $x = 0$ *or* $3x + 2 = 0$

i.e. $3x = -2$

$$x = -\frac{2}{3}.$$

So the roots of the equation are $x = 0$ or $x = -\frac{2}{3}.$

2. Solve $2x^2 + 9x + 4 = 0$.

Factorising $(2x + 1)(x + 4) = 0$.

Therefore $2x + 1 = 0$ *or* $x + 4 = 0$.

i.e. $2x = -1$ or $x = -4$

$$x = -\tfrac{1}{2}.$$

So the roots of the equation are $x = -\tfrac{1}{2}$ or $x = -4$.

Exercise 70

Solve (the equations are already in factor form).

1. $x(2x - 1) = 0$
2. $x(5x + 2) = 0$
3. $(5x + 1)(x - 2) = 0$
4. $(2x - 3)(x + 1) = 0$
5. $(x - 4)(5x + 6) = 0$
6. $(2x + 1)(x + 2) = 0$

7. $x(2 - 3x) = 0$
8. $(2x - 9)(x - 1) = 0$
9. $(5x - 4)(x + 3) = 0$
10. $(x - 1)(3x - 4) = 0$
11. $(7x + 1)(x + 1) = 0$
12. $(3 - 5x)(2 + x) = 0$

Exercise 71

Solve the following.

1. $2x^2 + 7x + 3 = 0$
2. $3x^2 + 5x + 2 = 0$
3. $5x^2 + 16x + 3 = 0$
4. $3x^2 - 7x + 2 = 0$
5. $5x^2 - 3x = 0$
6. $2x^2 - 5x + 3 = 0$
7. $5x^2 - 7x + 2 = 0$
8. $3x^2 + 19x + 6 = 0$

9. $2x^2 + 11x + 12 = 0$
10. $2x^2 + 9x = 0$
11. $2x^2 - 11x + 12 = 0$
12. $5x^2 - 22x + 8 = 0$
13. $2x^2 + 9x - 5 = 0$
14. $5x^2 - 17x - 12 = 0$
15. $5x - 4x^2 = 0$

Find the equation whose roots are -1 and $\tfrac{2}{5}$.

We can say that $x = -1$ or $x = \tfrac{2}{5}$

i.e. $x + 1 = 0$ or $5x = 2$

$$5x - 2 = 0.$$

Therefore $\quad (x + 1)(5x - 2) = 0$

i.e. $\quad 5x^2 + 3x - 2 = 0$.

Exercise 72

Find the equation whose roots are

1. $1, \frac{2}{3}$
2. $-2, \frac{1}{5}$
3. $\frac{2}{5}, 0$
4. $-\frac{1}{3}, -2$
5. $3, -\frac{4}{5}$
6. $0, -\frac{1}{7}$
7. $-5, \frac{1}{3}$
8. $\frac{1}{5}, 1$

Exercise 73

Solve the following equations, which need to be rearranged before factorising.

1. $x^2 + 9x = -14$
2. $x^2 = 5x - 6$
3. $x^2 = 144$
4. $x^2 - 8x = 30 - 7x$
5. $2x^2 = 3x$
6. $x^2 + x = 4 + x$
7. $\dfrac{x^2}{2} + x = 7\frac{1}{2}$
8. $x(3x - 1) = 1 + x^2$
9. $24 + 5x = x^2$

2.4 Quadratic equations (by formula)

All quadratic equations may be written in the form $ax^2 + bx + c = 0$. Quadratic equations which do not factorise may be solved using the formula

$$x = \frac{-b \pm \sqrt{b^2 - 4ac}}{2a}$$

where a = coefficient of x^2, b = coefficient of x, c = constant term.

Consider:

$x^2 + 5x + 3 = 0$ $\qquad a = 1, b = 5, c = 3$
$x^2 - 4x + 2 = 0$ $\qquad a = 1, b = -4, c = 2$
$x^2 - 6x - 6 = 0$ $\qquad a = 1, b = -6, c = -6$
$3x^2 - 7x + 3 = 0$ $\qquad a = 3, b = -7, c = 3.$

Exercise 74

Write down the value of a, b, and c in each of the following.

1. $x^2 + 6x + 6 = 0$
2. $x^2 - 9x + 17 = 0$
3. $x^2 + 3x - 2 = 0$
4. $x^2 - 10x + 6 = 0$
5. $x^2 + 4x + 2 = 0$
6. $x^2 - 5x + 5 = 0$
7. $x^2 - x - 5 = 0$
8. $x^2 + 3x - 6 = 0$
9. $x^2 - 3x - 9 = 0$
10. $x^2 - 6x + 3 = 0$
11. $x^2 - 4x - 6 = 0$
12. $x^2 + 3x + 1 = 0$

Using the formula

Example 1

Solve $x^2 + 5x + 3 = 0$.　　$a = 1, b = 5, c = 3$.

$$x = \frac{-b \pm \sqrt{b^2 - 4ac}}{2a}$$

$$x = \frac{-(5) \pm \sqrt{(5)^2 - 4(1)(3)}}{2(1)}$$

$$x = \frac{-5 \pm \sqrt{25 - 12}}{2}$$

$$x = \frac{-5 \pm \sqrt{13}}{2}$$

$$x = \frac{-5 \pm 3.606}{2}.$$

This gives $x = \dfrac{-5 + 3.606}{2}$　　or　　$x = \dfrac{-5 - 3.606}{2}$

$$x = \frac{-1.394}{2} \quad \text{or} \quad x = \frac{-8.606}{2}$$

$$x = -0.697 \quad \text{or} \quad x = -4.303.$$

So the roots of the equation $x^2 + 5x + 3 = 0$ are $x = -0.697$ or $x = -4.303$.

Example 2

Solve $x^2 - 4x + 2 = 0$.
$a = 1, b = -4, c = 2$.

$$x = \frac{-b \pm \sqrt{b^2 - 4ac}}{2a}$$

$$x = \frac{-(-4) \pm \sqrt{(-4)^2 - 4(1)(2)}}{2(1)}$$

$$x = \frac{+4 \pm \sqrt{16 - 8}}{2}$$

$$x = \frac{+4 \pm \sqrt{8}}{2}$$

$$x = \frac{+4 \pm 2.828}{2}.$$

This gives

$$x = \frac{+4 + 2.828}{2} \quad \text{or} \quad x = \frac{+4 - 2.828}{2}$$

$$x = \frac{6.828}{2} \quad \text{or} \quad x = \frac{1.172}{2}$$

$$x = 3.414 \quad \text{or} \quad x = 0.586.$$

Example 3

Solve $x^2 - 6x - 6 = 0$.

$a = 1, b = -6, c = -6$.

$$x = \frac{-b \pm \sqrt{b^2 - 4ac}}{2a}$$

$$x = \frac{-(-6) \pm \sqrt{(-6)^2 - 4(1)(-6)}}{2(1)}$$

$$x = \frac{+6 \pm \sqrt{36 + 24}}{2}$$

$$x = \frac{+6 \pm \sqrt{60}}{2}$$

$$x = \frac{+6 \pm 7.746}{2}.$$

This gives

$$x = \frac{+6 + 7.746}{2} \quad \text{or} \quad x = \frac{+6 - 7.746}{2}$$

$$x = \frac{13.746}{2} \quad \text{or} \quad x = \frac{-1.746}{2}$$

$$x = 6.873 \quad \text{or} \quad x = -0.873.$$

Exercise 75

Solve the following, using the quadratic formula.

1. $x^2 + 3x - 2 = 0$
2. $x^2 - 4x - 3 = 0$
3. $x^2 + 7x + 7 = 0$
4. $x^2 - 6x + 4 = 0$

5. $x^2 - 3x - 1 = 0$
6. $x^2 + 5x - 3 = 0$
7. $x^2 + 3x - 5 = 0$
8. $x^2 + 3x + 1 = 0$

9. $x^2 - 3x - 8 = 0$
10. $x^2 - 5x + 3 = 0$
11. $x^2 + 4x + 1 = 0$
12. $x^2 - 9x + 7 = 0$

Example 4

Solve $3x^2 - 7x + 3 = 0$. $a = 3, b = -7, c = 3$.

$$x = \frac{-b \pm \sqrt{b^2 - 4ac}}{2a}$$

$$x = \frac{-(-7) \pm \sqrt{(-7)^2 - 4(3)(3)}}{2(3)}$$

$$x = \frac{+7 \pm \sqrt{49 - 36}}{6}$$

$$x = \frac{+7 \pm \sqrt{13}}{6}$$

$$x = \frac{+7 \pm 3.606}{6}$$

This gives:

$$x = \frac{+7 + 3.606}{6} \quad \text{or} \quad x = \frac{+7 - 3.606}{6}$$

$$x = \frac{10.606}{6} \quad \text{or} \quad x = \frac{3.394}{6}$$

$x = 1.767$ or $x = 0.565$.

Exercise 76a

Solve the following, using the quadratic formula.

1. $2x^2 + 6x + 3 = 0$ 3. $5x^2 - x - 1 = 0$ 5. $2x^2 - 4x + 1 = 0$
2. $3x^2 - 5x - 4 = 0$ 4. $3x^2 + 2x - 3 = 0$ 6. $5x^2 + 8x + 2 = 0$

Special note There is no need to use the formula if

(a) there is *no constant term* in the quadratic equation; the equation will always factorise.

So that $x^2 + 3x = 0$ $5x^2 - 7x = 0$
            $x(x + 3) = 0$ $x(5x - 7) = 0$
           $x = 0$ or $x = -3$ $x = 0$ or $x = \frac{7}{5}$

(b) there is *no x term*. In this case there are two possibilities:

(i) $x^2 - 4 = 0$ $x^2 - 12 = 0$
      $x^2 = 4$ $x^2 = 12$
      $x = \sqrt{4}$ $x = \sqrt{12}$
      $x = \pm 2$ $x = \pm 3.464$

(ii)
$$x^2 + 9 = 0 \qquad\qquad x^2 + 8 = 0$$
$$x^2 = -9 \qquad\qquad\quad x^2 = -8$$
$$x = \sqrt{-9} \qquad\qquad x = \sqrt{-8}$$

The roots $\sqrt{-9}$ and $\sqrt{-8}$ have no meaning in the usual sense, and so expressions of this type cannot be solved (in the normal way). The roots are said to be 'imaginary'.

Exercise 76b

Solve the following. If the roots are imaginary, say so.

1. $x^2 - 7 = 0$
2. $x^2 + 4 = 0$
3. $5 - x^2 = 0$
4. $x^2 - 15 = 0$

5. $x^2 = 6$
6. $x^2 = -20$
7. $2x^2 - 8 = 0$
8. $-x^2 + 11 = 0$

9. $x^2 - 225 = 0$
10. $3x^2 = 60$
11. $-3 = x^2$
12. $x^2 - 3 = 0$

3   Equations – problems

The methods required for solving the equations in this chapter are the same as those already used, whether the equations be simple, simultaneous or quadratic.

The problem here is *to be able to form equations from the given information*, so that a solution may then be found.

3.1   Simple equations

Exercise 77

Form a simple equation from the information given and hence, solve the problem.

1. A triangle has sides $2x$ cm, $3x$ cm and $4x$ cm. Its perimeter is 18 cm. Find the lengths of the sides.

2. The sides of a triangle are given by $(2x+1)$ cm, $(x-1)$ cm and $(x+4)$ cm. Find the lengths of the sides when the perimeter is 16 cm.

3. An equilateral triangle of side $(x+1)$ cm has a perimeter of 24 cm. How long is each side?

4. A right-angled triangle with perimeter 30 cm, has sides of x cm, $(x+1)$ cm and $(x-7)$ cm. Find the lengths of these sides, and hence the area of the triangle.

5. A triangle has sides $(4x-1)$ cm, $(4x-5)$ cm and $(4x+2)$ cm, and a perimeter of 32 cm. Find x, and so determine the length of the shortest side.

6. The angles of a triangle are given by $x°$, $2x°$, and $3x°$. Find x, and hence the angles of the triangle.

7. A triangle has angles of $2x°$, $(3x+10)°$ and $(4x-10)°$. Find the angles. What sort of triangle is this?

8. A rectangle has a width of 5 cm, and a length of $(x+6)$ cm. If the area is 50 cm^2, form an equation to find x; hence find the length of the rectangle, and its perimeter.

9. The sides of a rectangle are given by $2x$ cm, $(x+4)$ cm, $(3x-3)$ cm

and $(2x+1)$ cm, and its perimeter is 26 cm. Find the value of x, and calculate the area of the rectangle.

10. The length of a rectangle is $(x+3)$ cm. and its width is 4 cm. If the perimeter is given by $(3x+4)$ cm, make an equation to find x, and hence find the actual perimeter.

11. The perimeter of a rectangle is $7x$, its width $(2x-3)$ and its length $(x+5)$. Find x, and hence the area of the rectangle.

12. The angles of a quadrilateral are given by $2x°$, $(x+30)°$, $(x+10)°$ and $(3x-30)°$. Find the size of each angle.

13. Triangle A has sides of x, $(x+3)$ and $(2x-4)$. Triangle B has sides of x, $(x+2)$ and $(4x-13)$.

If both triangles have the same perimeter, form an equation to find x. Hence find the lengths of the sides of both triangles, and their perimeters.

14. A rectangle is four times as long as it is wide, and has a perimeter of 60 cm. Find its length and its width.

15. The width of a rectangle is 3 cm less than its length. If its perimeter is 26 cm, find its length and width and hence its area.

16. The length of a rectangle is 4 cm more than the width, x cm. If the perimeter is given by $(6x-4)$ cm, form an equation in x, and find the actual perimeter.

Exercise 78

By forming simple equations, solve the following problems.

1. ABCD is a parallelogram.

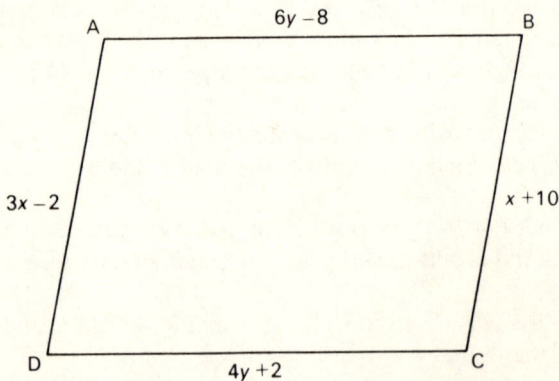

(a) Form an equation in x, and hence evaluate x.
(b) Form an equation in y, and hence evaluate y.
(c) Calculate the perimeter of the parallelogram.
2. PQR is an isosceles triangle, with PQ = PR.

Find the value of x, and hence the perimeter of the triangle.

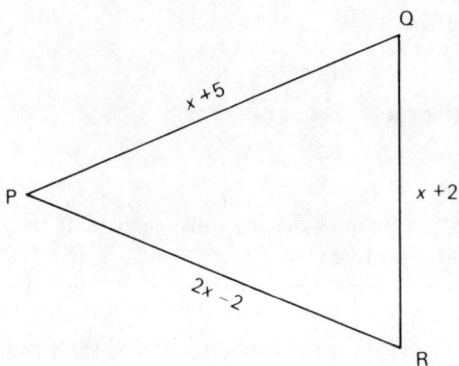

3. WXYZ is a right-angled trapezium, with a perimeter of 26 cm.

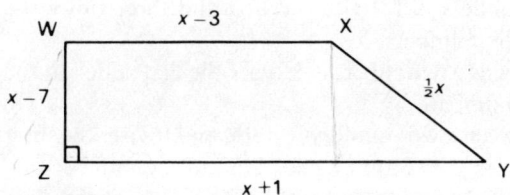

(a) Calculate the lengths of the sides.
(b) Find the area of the trapezium.
4. The figure represents a square prism, where ABCD and EFGH are squares.

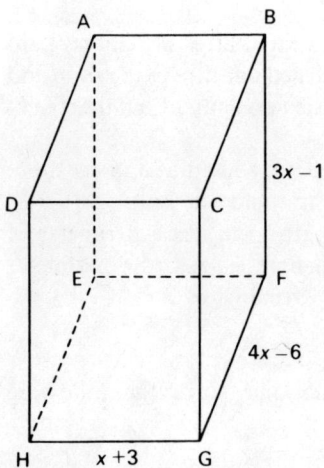

(a) Form an equation in x to find the dimensions of the shape.
(b) Calculate the total surface area of the prism.
(c) Find the volume of the prism.

3.2 Simultaneous equations

Exercise 79

In this exercise you will need to form two equations, each with two unknowns, in order to solve the problem.
1. The sum of two numbers is 7, and their difference is 1. Find the numbers.
2. The sum of two numbers is 11, and their difference is 5. Find the numbers.
3. The sum of two numbers is 1, the first number added to twice the second equals zero. Find the numbers.
4. The sum of two numbers is 1, twice the first and three times the second equals 5. Find the numbers.
5. The difference between two numbers is 5, twice the first added to the second is -8. Find the numbers.
6. The difference between two numbers is $1\frac{1}{2}$, and twice the first subtracted from four times the second equals 2. Find the numbers.
7. Three mathematics books and two science books cost 46 pence. Two mathematics books and three science books cost 44 pence. Find the cost of each type of book.
8. Five apples and two peaches cost 42 pence; two apples and four· peaches cost 52 pence. Find the cost of an apple and a peach.
9. The difference between two supplementary angles is 22°. What size is each angle?
10. A man bought six sparking plugs and three litres of oil, and paid £5.40. Finding he had too much oil, he returned one litre to the shop and was given four more sparking plugs as an exact exchange. Find the cost of one sparking plug and one litre of oil.
11. Five nuts and two bolts cost 24 pence. One less nut and six more bolts cost 64 pence. Find the cost of one nut and one bolt.
12. A rectangle with width x cm and length y cm has a perimeter of 12 cm. If the width is trebled and the length halved, the perimeter becomes 16 cm. Find the areas of the two rectangles.

Exercise 80

By forming pairs of simultaneous equations, solve the following problems.
1. KLMN is a rectangle.

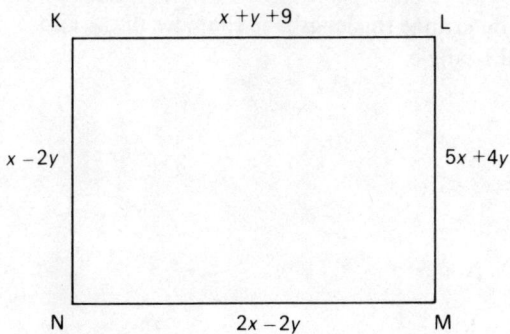

Rectangle KLMN with:
- KL (top): $x + y + 9$
- KN (left): $x - 2y$
- LM (right): $5x + 4y$
- NM (bottom): $2x - 2y$

(a) Find x and y.
(b) What is the perimeter of the rectangle?

2. ABCD is a rhombus.

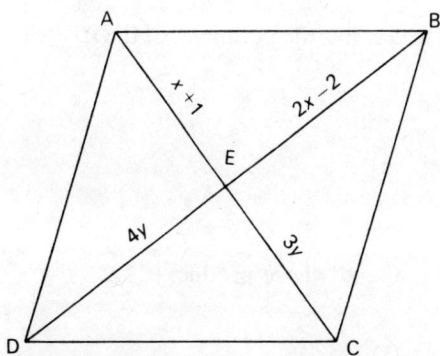

Rhombus ABCD with diagonals meeting at E:
- AE: $x + 1$
- EB: $2x - 2$
- AD side region: $4y$
- EC: $3y$

(a) Form equations to find x and y, and hence determine the lengths of the diagonals.
(b) Calculate the area of the rhombus.

3. DEFG is a kite, with a perimeter of 38 cm.

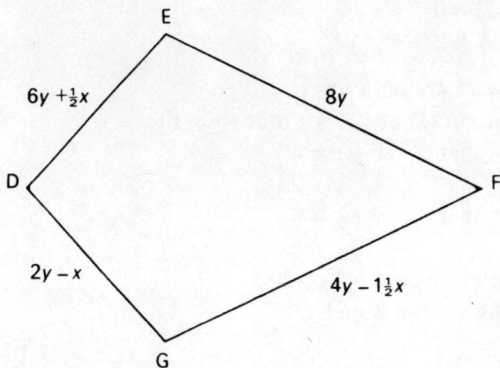

Kite DEFG with:
- DE: $6y + \frac{1}{2}x$
- EF: $8y$
- DG: $2y - x$
- GF: $4y - 1\frac{1}{2}x$

Find x and y, and hence determine the lengths of the sides of the kite.

4. PQR is an equilateral triangle.

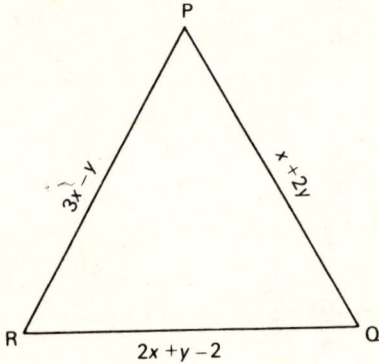

Find the values of x and y, and hence find the perimeter of the triangle.

3.3 Quadratic equations

Consider:

Example 1

Find two numbers whose sum is 8, and whose product is 12.

Let the numbers be x and y.
We can say that $x + y = 8$ ..... (1)
and $xy = 12$ ..... (2)
Re-arrange (1) $y = 8 - x$, and substitute in (2)
$$x(8 - x) = 12$$
$$8x - x^2 = 12$$
i.e. $x^2 - 8x + 12 = 0$ (a quadratic equation).
Factorising, $(x - 6)(x - 2) = 0$
i.e. $x = 6$ or $x = 2$.

Note that these two answers are *both for x*.
We now have to substitute in (1) or (2) in order to evaluate y.
Using (1), when $x = 6$ then $6 + y = 8$
$$y = 2$$
when $x = 2$ then $2 + y = 8$
$$y = 6.$$
Therefore the required solutions are 6,2 or 2,6.
In this case both solutions are the same!

Now consider:

64

Example 2

Find two numbers whose difference is 2, and whose product is 35.

Let the numbers be x and y.
We can say that $x - y = 2$ ..... (1)
and $xy = 35$ ..... (2)
Re-arrange (1) $y = x - 2$, and substitute in (2)
$$x(x - 2) = 35$$
$$x^2 - 2x = 35$$
i.e. $x^2 - 2x - 35 = 0$
Factorising, $(x - 7)(x + 5) = 0$
i.e. $x = 7$ or $x = -5$.
Substituting in (1), when $x = 7$ we have $7 - y = 2$
$$y = 5$$
when $x = -5$ we have $-5 - y = 2$
$$y = -7.$$
Therefore the required solutions are $7, 5$ or $-5, -7$.
In this case the two solutions are different, but both satisfy the conditions of the question.

Important note: remember that the solution of the quadratic is not the answer to the problem.

Exercise 81

Form quadratic equations in order to solve the following problems. (All examples will factorise.)

1. Find two numbers whose sum is 10, and whose product is 21.
2. Find two numbers whose difference is 3, and whose product is 18.
3. The sum of two numbers is 8, and their product is -20. What are the numbers?
4. The difference between two numbers is 4, and their product is -3. Find the numbers.
5. Two numbers have a product of 30, and a sum of -11. Find the numbers.
6. The sum of two numbers is 11, and the sum of their squares is 65. What are the numbers?
7. The length of a rectangle is 4 cm longer than the width. If the area is 12 cm^2, find the length and the width.
8. The length of a rectangle is 7 cm more than half the width. If the area is 120 cm^2, find the length and the width.
9. The area of a rectangle is 40 m^2, and its width is 3 m less than its length. Find the width and the length.

A—E

10. Two square rooms have a total floor area of 52 m². The smaller room is 2 m less each way than the larger. Find the floor dimensions of each room, and hence their separate areas.

11. The length and width of a rectangle measuring 4 cm by 7 cm are increased by the same amount, to form a rectangle with an area of 108 cm². By what amount were the sides increased?

12. The sum of the squares of two consecutive odd numbers is 202. What are the numbers?

13. The square of a number, less twice itself, equals 15. Find the number(s).

14. The squares of three consecutive even numbers added together total 116. Find the numbers.

15. A rectangle is 3 times as long as it is wide, and has an area of 75 m². Find the dimensions of the rectangle.

Exercise 82

Form quadratic equations to solve the following. (You may need to use the quadratic formula.)

1. ABC is a right-angled triangle, with an area of 10 cm².

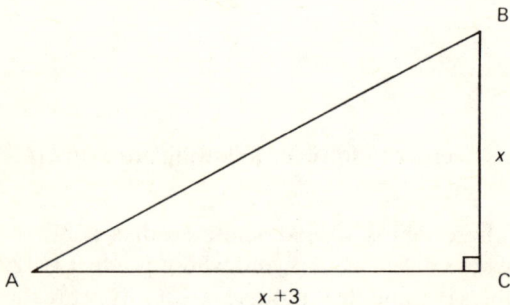

(a) Form an equation to evaluate x, and hence determine the lengths of AC and BC.
(b) Calculate the length of AB.
(Give your answers correct to 2 decimal places.)

2. XYZ is a right-angled triangle.
(a) Form an equation to find the value of x, and so obtain the lengths of XZ and YZ.
(b) Calculate the area of the triangle.

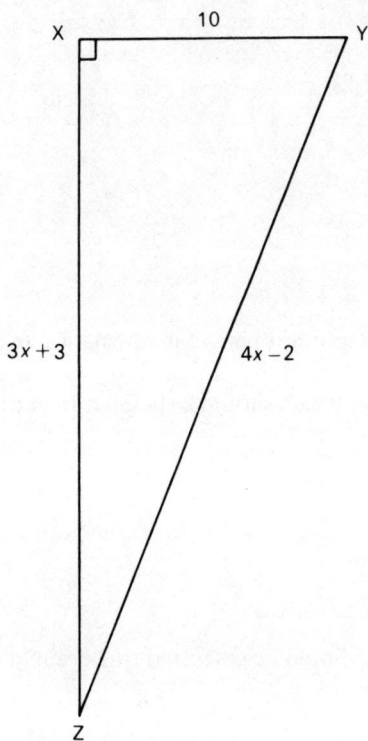

X — 10 — Y

$3x+3$ $4x-2$

Z

3. PQRS is a rhombus. $PR = 4x$, $SQ = 2x - 6$ and its area is 66 cm².

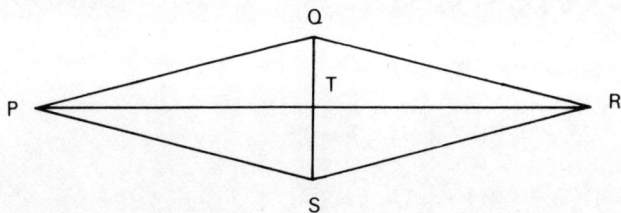

Q
T
P   R
S

(a) Form an equation to evaluate x, and hence work out the lengths of the diagonals.
(b) Calculate the length of the side of the rhombus.
(Give your answers correct to 2 decimal places.)

4 Formulae

The ability to handle formulae is an important part of mathematics. In this chapter you will be able to practise substituting into, and re-arranging, a variety of formulae some of which should be familiar to you from other aspects of your work.

4.1 Substitution

Exercise 83

Where decimals are involved, answers should be corrected to 2 decimal places.

The value of π should be taken as 3.14, when it is needed.

Find the value of:

1. $C = 2\pi r$,
   when (a) $r = 6$, (b) $r = 4.7$.
2. $A = \frac{1}{2}bh$,
   when (a) $b = 5$, $h = 8$, (b) $b = 2.4$, $h = 3.7$.
3. $V = lbh$,
   when (a) $l = 4$, $b = 7$, $h = 3$, (b) $l = 6.2$, $b = 3.3$, $h = 4.5$.
4. $V = \frac{1}{3}Ah$,
   when (a) $A = 25$, $h = 6$, (b) $A = 28$, $h = 4.5$.
5. $A = 2\pi rh$,
   when (a) $r = 10$, $h = 8$, (b) $r = 1.7$, $h = 1.4$.
6. $A = \pi rl$,
   when (a) $r = 2$, $l = 5$, (b) $r = 2.2$, $l = 1.9$.
7. $S = \dfrac{D}{T}$,
   when (a) $D = 150$, $T = 6$, (b) $D = 88$, $T = 5.5$.
8. $A = \left(\dfrac{x+y}{2}\right)h$,

when (a) $x = 7$, $y = 11$, $h = 3$, (b) $x = 1.8$, $y = 6.4$, $h = 2.5$.

9. $A = \dfrac{h}{2}(x + y)$,

when (a) $h = 10$, $x = 6$, $y = 3$, (b) $h = 4.3$, $x = 2$, $y = 3.6$.

10. $A = \frac{1}{2}ab \sin C$,

when (a) $a = 6$, $b = 8$, $C = 30°$, (b) $a = 4.5$, $b = 6.2$, $C = 18° \, 41'$.

11. $I = \dfrac{PRT}{100}$,

when (a) $P = 250$, $R = 6$, $T = 2$, (b) $P = 85$, $R = 7.5$, $T = 5.5$.

12. $A = \pi(R - r)(R + r)$,

when (a) $R = 9$, $r = 5$, (b) $R = 8.2$, $r = 4.1$.

13. $v = u + ft$,

when (a) $u = 0$, $f = 32$, $t = 6$, (b) $u = 7$, $f = 10$, $t = 8.5$.

14. $v = \sqrt{u^2 + 2fs}$,

when (a) $u = 20$, $f = 6$, $s = 100$, (b) $u = 5.2$, $f = 2$, $s = 7.5$.

15. $s = ut + \frac{1}{2}ft^2$,

when (a) $u = 3$, $t = 3$, $f = 8$, (b) $u = 0$, $t = 2.5$, $f = 32$.

16. $A = \pi r^2$,

when (a) $r = 10$, (b) $r = 7.2$.

17. $x = \sqrt{y^2 + z^2}$,

when (a) $y = 12$, $z = 5$, (b) $y = 3.2$, $z = 4.8$.

18. $V = \frac{4}{3}\pi r^3$,

when (a) $r = 2$, (b) $r = 1.2$.

19. $V = \frac{1}{3}\pi r^2 h$,

when (a) $r = 6$, $h = 5$, (b) $r = 4$, $h = 2.8$.

20. $A = \sqrt{s(s - a)(s - b)(s - c)}$, where $s = \frac{1}{2}(a + b + c)$,

when (a) $a = 5$, $b = 6$, $c = 5$, (b) $a = 5.4$, $b = 7.2$, $c = 6$.

21. $T = 2\pi \sqrt{\dfrac{l}{g}}$,

when (a) $l = 128$, $g = 32$, (b) $l = 25$, $g = 975$.

22. $a = \sqrt{b^2 + c^2 - 2bc \cos A}$,

when (a) $b = 6$, $c = 5$, $A = 60°$, (b) $b = 4.2$, $c = 5.5$, $A = 42° \, 25'$.

23. $A = P\left(1 + \dfrac{R}{100}\right)^N$,

when (a) $P = 200$, $R = 5$, $N = 8$, (b) $P = 175$, $R = 2.5$, $N = 6$.

24. $x = \dfrac{-b \pm \sqrt{b^2 - 4ac}}{2a}$,

when (a) $a = 2$, $b = 7$, $c = -4$, (b) $a = 1$, $b = -6$, $c = 3$.

4.2 Transposition (re-arrangement)

In the formula $A = \dfrac{h}{2}(x+y)$, the SUBJECT of the formula is A. If we need to make one of the other letters the subject, that is, h, x or y, then we have to re-arrange or TRANSPOSE the formula.

The processes involved in transposing formulae are the same as those used in the solution of equations, since a formula is just an equation with a special purpose.

Consider the following:

Example 1

Re-arrange $A = \dfrac{h}{2}(x+y)$ to make x the subject of the equation.

$$A = \dfrac{h}{2}(x+y)$$
$$2A = h(x+y) \quad \text{(Multiplying by 2)}$$
$$\dfrac{2A}{h} = x+y \quad \text{(Dividing by } h\text{)}$$
$$\dfrac{2A}{h} - y = x \quad \text{(Subtracting } y \text{ from both sides)}$$

Example 2

Make t the subject of the formula $v = u + ft$.

$$v = u + ft$$
$$v - u = ft \quad \text{(Subtracting } u \text{ from both sides)}$$
$$\dfrac{v-u}{f} = t \quad \text{(Dividing by } f\text{)}$$

Example 3

Transpose $V = \frac{1}{3}\pi r^2 h$ to find a formula for r.

$$V = \tfrac{1}{3}\pi r^2 h$$
$$3V = \pi r^2 h \quad \text{(Multiplying by 3)}$$
$$\dfrac{3V}{\pi h} = r^2 \quad \text{(Dividing by } \pi h\text{)}$$

$$\sqrt{\dfrac{3V}{\pi h}} = r \quad \text{(Taking the square root)}$$

Example 4

Make l the subject of the formula $T = 2\pi \sqrt{\dfrac{l}{g}}$

$T = 2\pi \sqrt{\dfrac{l}{g}}$

$T^2 = 4\pi^2 \dfrac{l}{g}$ (Squaring)

$gT^2 = 4\pi^2 l$ (Multiplying by g)

$\dfrac{gT^2}{4\pi^2} = l$ (Dividing by $4\pi^2$)

Exercise 84

Re-arrange the following as indicated.
1. $C = 2\pi r$, make r the subject of the formula.
2. $A = \frac{1}{2}bh$, re-arrange to make a formula for b.
3. $A = \pi r l$, transpose for l.
4. $S = \dfrac{D}{T}$, make T the subject of the formula.
5. $A = \frac{1}{2}ab \sin C$, transpose to obtain an expression for $\sin C$.
6. Re-arrange $v^2 = u^2 + 2fs$ to obtain a formula for s.
7. Re-arrange $v^2 = u^2 + 2fs$ to make a formula for u.
8. Transpose $I = \dfrac{PRT}{100}$, to find an expression for T.
9. Make z the subject of the formula $x^2 = y^2 + z^2$.
10. Re-arrange $A = \pi r^2$, so that r becomes the subject.
11. Transpose $T = 2\pi \sqrt{\dfrac{l}{g}}$, to make a formula for g.
12. Make R the subject of $A = \pi(R - r)(R + r)$.

Exercise 85

Re-arrange and evaluate. (Give your answers correct to 2 decimal places where necessary, take $\pi = 3.14$)
1. Make H the subject of $V = LBH$, and evaluate when $V = 270$, $L = 9$, $B = 6$.
2. If $S = \dfrac{D}{T}$, re-arrange for D.
What is the value of D when $S = 55$, $T = 3.25$?
3. Re-arrange $V = \frac{1}{3}Ah$ for A, and calculate the value of A when $V = 128$, $h = 16$.

4. If $v = u + ft$, re-arrange for f, and evaluate when $v = 65.5$, $u = 6$, $t = 7$.

5. From $x^2 = y^2 + z^2$ make a formula for y, and calculate the value of y when $x = 8.06$, and $z = 4$.

6. If $A = \dfrac{h}{2}(x + y)$, make h the subject. Evaluate when $A = 83.7$, $x = 12$, $y = 15$.

7. If $I = \dfrac{PRT}{100}$, find P when $R = 7$, $T = 5$, $I = 210$.

8. From $v^2 = u^2 + 2fs$, calculate s if $v = 7.5$, $u = 3.2$, $f = 8.8$.

9. If $T = 2\pi \sqrt{\dfrac{l}{g}}$, find the value of l when $T = 2.5$, $g = 975$.

10. Transpose $a^2 = b^2 + c^2 - 2bc \cos A$ to give a formula for $\cos A$, and hence determine $\angle A$ when $a = 9$, $b = 7$, $c = 8$.

11. Re-arrange $V = \frac{4}{3}\pi r^3$, to find an expression for r. Evaluate when $V = 500$.

12. $A = \pi(R - r)(R + r)$. Find r when $R = 12$ and $A = 200$.

5 Indices

5.1 Laws of indices

1 Multiplication

$$x^a \times x^b = x^{a+b}$$

Examples

(i) $x^6 \times x^3 = x^{6+3}$
$\qquad \quad = x^9$

(ii) $6^4 \times 6^3 = 6^{4+3}$
$\qquad \qquad = 6^7$

(iii) $2^2 \times 2^3 \times 2^5 = 2^{2+3+5}$
$\qquad \qquad \quad = 2^{10}$

(iv) $y^{-7} \times y^4 = y^{-7+4}$
$\qquad \qquad = y^{-3}$

2 Division

$$x^a \div x^b = x^{a-b}$$

Examples

(i) $x^5 \div x^3 = x^{5-3}$
$\qquad \quad = x^2$

(ii) $\dfrac{10^8}{10^3} = 10^{8-3}$
$\qquad \quad = 10^5$

(iii) $y^6 \div y^{-2} = y^{6-(-2)}$
$\qquad \qquad = y^8$

(iv) $3^6 \div 3^4 \times 3^2 = 3^{6-4+2}$
$\qquad \qquad \qquad = 3^4$

3 Powers

$$(x^a)^b = x^{ab}$$

Examples

(i) $(x^3)^2 = x^{3 \times 2}$
$\qquad = x^6$

(ii) $(4^4)^3 = 4^{4 \times 3}$
$\qquad = 4^{12}$

(iii) $(3x^2)^4 = 3^{1 \times 4} x^{2 \times 4}$
$\qquad = 3^4 x^8$
$\qquad = 81x^8$

(iv) $\left(\dfrac{x^4}{2y^3} \right)^3 = \dfrac{x^{4 \times 3}}{2^{1 \times 3} \, y^{3 \times 3}}$

$\qquad\qquad = \dfrac{x^{12}}{8y^9}$

4 (a) Roots

$$x^{1/a} = \sqrt[a]{x}$$

Examples

(i) $x^{1/3} = \sqrt[3]{x}$

(ii) $16^{1/2} = \sqrt[2]{16}$
$\qquad = 4$

(iii) $(8y^9)^{1/3} = \sqrt[3]{8y^9}$
$\qquad = 2y^3$

(iv) $xy^{1/5} = x\sqrt[5]{y}$

(b) Power/root combined

$$x^{b/a} = \sqrt[a]{x^b}$$

Examples

(i) $x^{3/4} = \sqrt[4]{x^3}$

(ii) $8^{2/3} = \sqrt[3]{8^2}$ *or* $(8^{1/3})^2$

$\qquad\qquad = \sqrt[3]{64} \qquad\qquad\quad (2)^2$
$\qquad\qquad = 4 \qquad\qquad\qquad\quad 4$

5 Reciprocal

$$x^{-a} = \dfrac{1}{x^a}$$

Examples

(i) $y^{-4} = \dfrac{1}{y^4}$ (ii) $5x^{-2} = \dfrac{5}{x^2}$

(iii) $(5x)^{-2} = \dfrac{1}{(5x)^2}$ (iv) $x^2 y^{-3} z^{-2} = \dfrac{x^2}{y^3 z^2}$

$\qquad\qquad = \dfrac{1}{25x^2}$

(v) $\dfrac{1}{5^{-4}} = (5^{-4})^{-1}$ (vi) $25^{-3/2} = \dfrac{1}{25^{3/2}}$

$\qquad\quad = 5^4$ $\qquad\qquad = \dfrac{1}{(25^{1/2})^3}$

$\qquad\quad = 625$ $\qquad\qquad = \dfrac{1}{5^3}$

$\qquad\qquad\qquad\qquad\qquad = \dfrac{1}{125}$

6 Zero index

$$\boxed{x^0 = 1}$$

Any expression to the power zero $= 1$.
Examples

(i) $6y^0 = 6 \times y^0$ (ii) $(6y)^0 = 6^0 \times y^0$
$\quad\; = 6 \times 1$ $\qquad\quad = 1 \times 1$
$\quad\; = 6$ $\qquad\quad = 1$

(iii) $(27)^0 = 1$ (iv) $(5x^{3/4})^0 = 1$

Exercise 86

Simplify the following (leave numerical answers in index form).

1. $y^2 \times y^3$ 5. $a^{-3} \times a^6$ 9. $5^2 \times 5^3 \times 5$
2. $5^4 \times 5^3$ 6. $2^3 \times 2^0$ 10. $2^{-2} \times 2^3 \times 2^4$
3. $x^6 \times x$ 7. $x^2 y \times xy^{-4}$ 11. $4^0 \times 4^6 \times 4$
4. $2y^2 \times 6y^2$ 8. $3^5 \times 3^7$ 12. $5^3 \times 2^0$

13. $x^7 \div x^3$

14. $2^9 \div 2^2$

15. $2^2 \div 2^9$

16. $m^3 \div m^3$

17. $\dfrac{4^3}{4^5}$

18. $\dfrac{2x^5}{4x^3}$

19. $\dfrac{2^{-3}}{2^{-3}}$

20. $\dfrac{5^6}{5^0}$

21. $a^6 \div a^2 \times a^3$

22. $2^{-5} \times 2^0 \div 2^{-5}$

23. $\dfrac{3^3 \times 3^4}{3^2 \times 3^6}$

24. $\dfrac{l^2 \times l^3}{l \times l^2}$

25. $(5^2)^2$

26. $(3x^4)^3$

27. $(2^4)^3$

28. $(6x^0)^2$

29. $(x^2 y^3)^5$

30. $(2ab^2)^3$

31. $\left(\dfrac{2x^2}{3y}\right)^4$

32. $(x^3 yz^2)^0$

33. $(2^2)^3 \times (2^2)^2$

34. $(x^2 y)^4 \times (2xy^4)^2$

35. $(4^2)^5 \div (4^3)^2$

36. $\dfrac{(5^6)^2 \times (5^2)^2}{(5^3)^2 \times (5^2)^4}$

Exercise 87

Evaluate the following.

1. 7^0

2. $9^{\frac{1}{2}}$

3. $16^{\frac{1}{4}}$

4. $8^{\frac{1}{3}}$

5. 8^{-1}

6. 2^{-3}

7. 3^{-2}

8. 5^{-2}

9. $(\tfrac{1}{2})^{-1}$

10. $(\tfrac{3}{4})^{-2}$

11. $(2^3)^{-2}$

12. $(3^{-4})^0$

13. $8^{\frac{2}{3}}$

14. $16^{\frac{3}{4}}$

15. $9^{\frac{3}{2}}$

16. $\left(\dfrac{1}{4}\right)^{\frac{5}{2}}$

17. $4^{-\frac{1}{2}}$

18. $27^{-\frac{2}{3}}$

19. $81^{-\frac{1}{4}}$

20. $125^{-\frac{1}{3}}$

21. $\left(\dfrac{16}{25}\right)^{\frac{1}{2}}$

22. $\left(\dfrac{9}{4}\right)^{-\frac{1}{2}}$

23. $(4^{\frac{1}{2}})^{-3}$

24. $(64^{\frac{1}{2}})^{\frac{1}{3}}$

25. $\dfrac{1}{2^{-4}}$

26. $(16x^8)^{\frac{1}{4}}$

27. 5×5^{-2}

28. $4^{\frac{1}{2}} \times 4^3$

29. $125^{\frac{1}{3}} \times 100^{\frac{1}{2}}$

30. $(36)^{-\frac{1}{2}} \times (27)^{\frac{2}{3}}$

31. $3^{2x} \times 3^x$

32. $(4^{6x})^{\frac{1}{3}}$

Indices in equations

Consider:

Example 1

Find the value of x if $2^x = 16$.

$$2^x = 16$$
$$2^x = 2^4$$

therefore $x = 4$.

Example 2

Find the value of x if $5 \times 25^x = 5^{3x}$

$$5 \times 25^x = 5^{3x}$$
$$5 \times (5^2)^x = 5^{3x}$$
$$5 \times 5^{2x} = 5^{3x}$$
$$5^{2x+1} = 5^{3x}$$

therefore $2x + 1 = 3x$
i.e. $x = 1.$

In order to solve this kind of equation, all the terms have to be expressed in the SAME BASE, so that the indices can then be compared. As with all equations, the result can be checked by substituting back into the original statement.

Exercise 88

Find the value of x in the following equations.

1. $4^x = 64$
2. $2^{3x} = 64$
3. $1000^x = 10$
4. $3^x = \frac{1}{9}$

5. $4 \times 2^x = 32$
6. $25^x \times 5^{-1} = 25^{\frac{3}{2}}$
7. $2^{2x} = \frac{1}{32}$
8. $3^{3x} \times 9 = 27^{-1}$

9. $4^{-2x} \times 16 = 8$
10. $25^{\frac{1}{x}} = 5$
11. $1000 = 10^x$
12. $(3^{2x})^2 \times 81^{\frac{1}{2}} = 9^3$

5.2 Standard form (for *very large* and *very small* numbers)

Indices provide a useful means of writing very large and very small numbers.

For example, the mass of the Earth is

5 975 000 000 000 000 000 000 000 kg,

which may be written more conveniently as 5.975×10^{24} kg.
Again, the number 0.000 000 007 may be written as 7.0×10^{-9}.

This method of writing numbers, so as to avoid long strings of zero's, is called STANDARD FORM (can also be called INDEX NOTATION).

When using standard form, numbers are written as $A \times 10^n$, where A lies between 1 and 10, and n is an integer.
(Note: n will be positive for very large numbers and negative for very small numbers.)

Exercise 89

Express the following in standard form.

1. 52000
2. 0.00000495
3. 65000000000
4. 0.00815
5. 990000000
6. 0.0000000004
7. 2500000
8. 0.0073
9. 512000000000000
10. 0.00000000000085

Exercise 90

Write the following in normal notation.

1. 1.2×10^3
2. 6.26×10^{-5}
3. 9.8×10^{10}
4. 3.7×10^{-2}
5. 5.2×10^6
6. 1.1×10^{-9}
7. 4.17×10^{-5}
8. 1.0×10^5
9. 8.95×10^{12}
10. 2.35×10^{-8}

Exercise 91

1. The distances, in kilometres, of the planets in our solar system from the sun are (approximately) as follows:

Mercury 58000000 Venus 108000000
Earth 150000000 Mars 228000000
Jupiter 778000000 Saturn 1427000000
Uranus 2870000000 Neptune 4497000000
Pluto 5910000000.

Express these distances in standard form.

2. Write the populations of the following countries in normal notation.
Japan 1.05×10^8 India 5.5×10^8 China 7.32×10^8 United Kingdom 5.5×10^7 United States of America 2.03×10^8 Soviet Union 2.42×10^8

3. (a) A single bacterium may weigh 2×10^{-12} g. Express this as an ordinary number.
(b) The diameter of the hydrogen nucleus is 1.0×10^{-15} m. Write this in ordinary notation.
(c) The outer membrane of a cell is approximately 0.000008 mm thick. Write this in standard form.
(d) The diameter of a body cell is 0.001 cm. Express this in standard form.

4. (a) The age of the Earth is 4.7×10^9 years. Write this as an ordinary number.
(b) The remaining life of the Sun is estimated to be 5000 million years. Express this in standard form.

(c) An average human body contains 5.0×10^{13} cells. How many is this as an ordinary number?

(d) When seeing violet light the retina of the eye receives about 1 000 000 000 000 000 vibrations per second. Write this in standard form.

Calculations using standard form

Example 1

$4.2 \times 10^6 \times 3.0 \times 10^4$
$4.2 \times 3.0 \times 10^6 \times 10^4$
12.6×10^{10}
i.e. 1.26×10^{11}.

Example 2

$5.26 \times 10^8 \div 7.28 \times 10^3$

$$\frac{5.26 \times 10^8}{7.28 \times 10^3}$$

0.7225×10^5
i.e. 7.225×10^4.

Example 3

$6.4 \times 10^{-3} \times 3.5 \times 10^{-4}$
$6.4 \times 3.5 \times 10^{-3} \times 10^{-4}$
22.4×10^{-7}
i.e. 2.24×10^{-6}.

Exercise 92

In the first six questions, simplify and write the answer in standard form.

1. $2.8 \times 10^5 \times 1.6 \times 10^4$
2. $5.6 \times 10^{10} \div 3.2 \times 10^4$
3. $1.2 \times 10^6 \times 9.7 \times 10^{-4}$
4. $3.35 \times 10^{-6} \div 2.08 \times 10^7$
5. $1.25 \times 10^{-3} \times 5.5 \times 10^{-9}$
6. $6.7 \times 10^4 \div 9.9 \times 10^{-5}$

7. One *light year* is the distance travelled by light in one year, and is equal to 9.46×10^{12} km.
(a) How far distant in kilometres are:
 (i) The Pleiades, 410 light years away.
 (ii) Alpha Centauri, 4.29 light years away.
 (iii) The Orion Nebula, 1.6×10^3 light years away.
 (iv) The Great Spiral Nebula in Andromeda, 2.0×10^6 light years away?
(b) How many light years away are:
 (i) Barnards Star, 5.648×10^{13} km.
 (ii) Sirius, 8.23×10^{13} km.
 (iii) Antares, 1.012×10^{15} km.
 (iv) The Crab Nebula, 4.73×10^{16} km?

8. Visible radiations are measured in ÅNGSTRÖM UNITS, equal to 1.0×10^{-8} cm.
(a) The wavelength of red colours is 7.0×10^3 ångström.
(b) The wavelength of deep violet is 3.5×10^3 ångström.
Write these wavelengths as centimetres, expressed in standard form.

6 Examination questions

The majority of questions in this chapter are taken from past Mathematics papers of the CSE examining boards. Exercise 93 contains multiple-choice questions, exercise 94 contains questions taken from first papers, and exercise 95 contains questions from second papers.

Exercise 93

For each of the following questions choose the correct answer from the choices listed.

1. Solve the equation $17x = -51$.
   (a) $x = -68$ (b) $x = -3$ (c) $x = -\frac{1}{3}$
   (d) $x = \frac{1}{3}$ (e) $x = 3$ (*NWSSEB*)

2. $\frac{1}{a} + \frac{1}{b}$ is equal to

   (a) $\frac{1}{a+b}$ (b) $\frac{2}{a+b}$ (c) $\frac{a+b}{ab}$ (d) $\frac{a+b}{2}$ (e) $\frac{1}{ab}$ (*WMEB*)

3. When $x = -3$, the expression $4 - x^2$ has the value
   (a) -5 (b) -2 (c) 10   (d) 13   (*Met. REB*)

4. If $\frac{x}{3} = \frac{5}{6}$, then x has the value

   (a) 2   (b) $2\frac{1}{2}$ (c) 5   (d) 10   (*Met. REB*)

5. If $N = 4 \times 10^{-2}$, then N^2 in standard form is
   (a) 16×10^{-2} (b) 16×10^{-4} (c) 16×10^4 (d) 1.6×10^{-3}
                                                          (*Met. REB*)

6. A solution to the equation $x^2 - 2x - 3 = 0$ is
   (a) $x = 1$ (b) $x = 2$ (c) $x = 3$ (d) $x = 4$ (*Met. REB*)

7. If $A = \frac{B+C}{2}$, then C is

   (a) $2A + B$ (b) $A - \frac{B}{2}$ (c) $A + \frac{B}{2}$ (d) $2A - B$ (*Met. REB*)

8. 2.15×10^{-5} equals
   (a) 0.00000215   (b) 0.00215   (c) 0.0000215   (d) 0.43
   (e) -107.5 (*WMEB*)

9. If $x = -3$ then the value of $2x^2 - 3x - 9$ is
   (a) 0   (b) 18   (c) -18 (d) 24   (e) none of these
   (WMEB)

10. $x^2 + 4$ is equal to
    (a) $(x+2)^2$ (b) $(x+2)(x-2)$ (c) $(x-2)^2$ (d) $2(x+2)$
    (e) none of these   (WMEB)

11. When $3x + 2y = 9$
    (a) the value of x, when $y = -3$ is: 5; $1\frac{2}{3}$; 1; $\frac{1}{3}$; 0.
    (b) the value of y, when $x = \frac{1}{3}$ is: 3; 4; 5; 6; none of these.
    (WYLEB)

12. If $f = a + gt$, then t is equal to
    (a) $\dfrac{f-a}{g}$ (b) $\dfrac{g}{f-a}$ (c) $fg - ag$ (d) $f - a - g$
    (e) $f + a + g$ (WMEB)

13. The value of 3.14×10^3 is
    314; 3140; 31400; 314000; 3314   (WYLEB)

14. $(x+6)(x-2)$ is equal to
    (a) $x^2 - 12$ (b) $2x + 4$ (c) $x^2 + 4x - 12$ (d) $x^2 + 8x - 12$
    (e) $x^2 - 6x + 4$ (WMEB)

15. The value of x which satisfies the equation $3x - 2 = x + 4$ is
    $x = 2$; $x = 3$; $x = \frac{1}{2}$; $x = 1\frac{1}{2}$; $x = 4$ (WMEB)

16. The value of $2^3 \times 2^2$ is: 24; 10; 36; 12; 32   (WMEB)

17. Simplify $k^3 + k^3$ k^6; k^9; $2k^3$; $2k^6$; $2k^9$ (NWSSEB)

18. The value of x which satisfies the equation $3x + 4 = x + 5$ is
    2; -1; $\frac{1}{2}$; $2\frac{1}{4}$; $4\frac{1}{2}$ (WMEB)

19. $x^2 + 4x - 12$ equals
    (a) $(x+6)(x-2)$ (b) $(x-6)(x+2)$ (c) $(x-3)(x-4)$
    (d) $(x-6)(x-2)$ (e) $(x+4)(x-3)$ (WMEB)

20. Simplify $(bc^3)^2$. $2bc^3$; bc^5; b^2c^5; b^2c^6; b^2c^9 (NWSSEB)

21. Evaluate 4^{-2}. -16; -8; $-\frac{1}{2}$; $\frac{1}{16}$; $\frac{1}{2}$ (NWSSEB)

22. The square root of $16x^{16}$ is: $4x^4$; $8x^4$; $4x^8$; $8x^8$; $4x^{16}$ (NWSSEB)

23. Simplify $\dfrac{8p^4}{2p}$ $4p^3$; $6p^3$; $4p^4$; $6p^4$; $16p^5$ (NWSSEB)

24. If $1.834 \times 10^n = 0.01834$ the value of n will be
    -3: -2; 0; 2; 3   (NWSSEB)

25. Solve the equation $\dfrac{x}{6} = \dfrac{2}{3}$ $-5\frac{1}{3}$; $\frac{1}{9}$; $\frac{1}{4}$; 4; 9   (NWSSEB)

26. Solve the equation $2x - 3 = 0$ $-\frac{3}{2}$; $-\frac{2}{3}$; $\frac{2}{3}$; $\frac{3}{2}$; 1   (NWSSEB)

27. What is the sum of $a + b$ and $a - b$
    $2a$; $2b$; $2a - 2b$; $2a + 2b$; $a^2 - b^2$ (NWSSEB)

28. What is the product of $(2x+3)(x-2)$
    $2x^2 - 6$; $2x^2 + x + 6$; $2x^2 + x - 6$; $2x^2 - x + 6$; $2x^2 - x - 6$ (NWSSEB)

81

A—F

29. If $3^x = 9$, then the value of x is

$\frac{1}{3}; -2; 2; 3; 27$ (*WYLEB*)

30. If $a = -1$ and $b = 3$, then the value of $\dfrac{a+2b}{b-a}$ is

$-3; 1.25; 2; 3; 3.5$ (*WYLEB*)

Exercise 94

1. Make f the subject of the formula $s = ut + \frac{1}{2}ft^2$.
2. If $\quad 13x + y = 33$
   and $6x + 2y = 26 \qquad$ find the values of x and y. (*SREB*)
3. Solve the equation $\dfrac{x}{3} + 2 = \dfrac{x}{4} + 3$. (*WMEB*)
4. Expand (a) $5x(2 - x^2)\qquad$ (b) $(2 + x)(5 - x)$.
   Factorise (c) $2x^3 - 10x^2 + 6x \qquad$ (d) $x^2 - 7x + 12$
   (e) $xy - xz + x^2$.
5. Solve the following equations:

   (a) $5x - 2 = 23 \qquad$ (b) $4(2 + x) = 16 \qquad$ (c) $\dfrac{x}{2} - \dfrac{x}{5} = \dfrac{3}{10}$

   (d) $4x + 2y = 16$
   $\quad\ 2x - 3y = 0$

6. When $a = 3$, $b = -2$, $c = 4$, $d = -5$, find the value of:
   (a) $a + b + c - d \qquad$ (b) $5a + 3d \qquad$ (c) $a^3 + d^2$

   (d) $c^{1/2} \qquad$ (e) b^{-2}.
7. Remove the brackets and simplify $\frac{1}{2}(2x + 6) + 3(x - 1)$. (*Met. REB*)
8. If $x = -2$ and $y = -3$, find the value of $x + y$. $\qquad$ (*Met. REB*)
9. In parts (a) and (b) give your answers in the simplest form.
   (a) Add $(x + 1)$ to $(x - 1)$. $\qquad$ (b) Multiply $(x + 1)$ by $(x - 1)$.

   (c) Evaluate $\dfrac{x+1}{x-1}$, when $x = 0$. $\qquad$ (d) If $\dfrac{x+1}{x-1} = 2$, evaluate x.

   (e) If $\dfrac{x+1}{x-1} = y$, express x in terms of y. (*SREB*)
10. Find the value of $x^2 - 5x - 6$, when $x = -3$. $\qquad$ (*WMEB*)
11. Solve the equations $\quad y = 3x + 4$
    $\qquad\qquad\qquad\qquad 3x = 2y - 5$. $\qquad$ (*WMEB*)
12. If $x = -2$ and $y = -3$, evaluate $(5 + x + y)^2 - (x + y) + xy + 9$
    $\qquad$ (*WMEB*)
13. If $x = 13 + 5y$, obtain a formula which gives y in terms of x.
    $\qquad$ (*WMEB*)
14. Solve the simultaneous equations $3a + 2b = 11$ and $a - 2b = 1$.
    $\qquad$ (*WMEB*)

15. Find x if: $2(5+x)-(2-x)=14$. $\hspace{2cm}$ (*WMEB*)

16. Express as a single fraction: $\dfrac{1}{x}-\dfrac{2}{(3+2x)}$. $\hspace{1cm}$ (*WMEB*)

17. Multiply out $(4x+3)(2x-5)$. $\hspace{2cm}$ (*WMEB*)

18. $p=-5$ and $q=3$. Find the value of:
   (a) $2p+3q$ $\hspace{0.5cm}$ (b) $q-p$ $\hspace{0.5cm}$ (c) p^2q. $\hspace{1cm}$ (*EMREB*)

19. Simplify (a) $x+2x+4x$ $\hspace{0.5cm}$ (b) $6x^2 \div 2x$ $\hspace{0.5cm}$ (c) $\sqrt{x^4}$. $\hspace{0.2cm}$ (*EMREB*)

20. If $x=2y$, what is the value of:
   (a) x, when $y=2$ $\hspace{0.5cm}$ (b) y, when $x=2$ $\hspace{0.5cm}$ (c) y, when $x+y=3$
   (d) $\dfrac{x}{y}$. $\hspace{4cm}$ (*WYLEB*)

21. If 5 is the average of the numbers 1, 2, 4, 6, x, calculate the value of x. $\hspace{3cm}$ (*Met. REB*)

22. If $\dfrac{x}{x+1}=\dfrac{5}{7}$, find the value of x. $\hspace{2cm}$ (*Met. REB*)

23. If $p=q\sqrt{k}$, find k when $p=32$ and $q=8$. $\hspace{1cm}$ (*Met. REB*)

24. Simplify $(4x-3y)-(3x+4y)$. $\hspace{2cm}$ (*Met. REB*)

25. (a) $32 \times 16 = 2^x$, calculate x.
   (b) $(2^4)^2$ as a power of 2 is? $\hspace{2cm}$ (*Met. REB*)

26. Find the value of xy^2+x^2y, when $x=-3$ and $y=4$. $\hspace{0.3cm}$ (*Met. REB*)

27. Given that $x-2y=6$ and $y-2x=6$, find the value of $x-y$. $\hspace{4cm}$ (*Met. REB*)

28. Simplify $3(4-x)-2(8-3x)$. $\hspace{2cm}$ (*WMEB*)

29. Factorise: (a) $3abc-9acd$ $\hspace{0.5cm}$ (b) $a(b-c)+(b-c)$. $\hspace{0.5cm}$ (*WJEC*)

30. Solve the simultaneous equations $2a+b=6$
   $$a-b=3.$$ $\hspace{4cm}$ (*WJEC*)

31. Simplify: (a) $9a+3a$ $\hspace{0.5cm}$ (b) $-9a+3a$ $\hspace{0.5cm}$ (c) $-9a \div (-3a)$
   (d) $-9a \times 3a$. $\hspace{3cm}$ (*WJEC*)

32. Solve $5x^2-2x-3=0$.

33. Given that $r=7$, $\pi=\frac{22}{7}$, and $h=9$, find the value of:
   (a) $2\pi r$ $\hspace{0.5cm}$ (b) πr^2 $\hspace{0.5cm}$ (c) $2\pi rh$ $\hspace{0.5cm}$ (d) $\frac{1}{3}\pi r^2 h$. $\hspace{1cm}$ (*WJEC*)

34. $A=\frac{1}{2}(c+d)h$
   (a) Find A if $c=7$, $d=5$ and $h=6$
   (b) Find h if $A=20$, $c=3$ and $d=2$. $\hspace{1.5cm}$ (*WJEC*)

35. Solve the equation: $\dfrac{3-2x}{3}-\dfrac{3-4x}{2}=\dfrac{13}{6}$. $\hspace{1cm}$ (*WJEC*)

36. If $a=-2$, $b=0$, $c=5$, find the value of $c^2+2a-abc$.

37. Solve $3x-2y=11$
   $$2x+y=12.$$

38. $x^2=16$. Write down one value for x. $\hspace{1.5cm}$ (*Met. REB*)

39. Solve the equation $\dfrac{x}{3}+\dfrac{x}{4}=7$. $\hspace{1.5cm}$ (*Met. REB*)

40. The angles of an isosceles triangle, measured in degrees, are x, $2x$, and $2x$. Find the value of x. (WMEB)

41. The surface area of a sphere is given by the formula: area $= 4\pi r^2$, where r is the radius of the sphere. Calculate the surface area of a sphere of radius 7 cm. Give your answer in cm². (Take $\pi = 22/7$)

42. Express 4.08×10^3 in normal notation.

43. If $A = 2\pi rh + 2\pi r^2$, express h in terms of A, π and r.

44. The angles of a triangle are $x°$, $3x°$, and $5x°$.
    Calculate the largest angle. (Met. REB)

45. Solve the equation $x^2 - 5x - 14 = 0$. (WMEB)

46. Factorise the expression: $8a^2 - 6ab + 2a$.

47. If $w = 3$, $x = -1$, $y = 4$, $z = -2$, evaluate $(x+z)^2 + wy$.

48. Solve $x^2 - 8x + 15 = 0$.

49. The third angle of an isosceles triangle is three times as big as either of the two equal angles. Find the size of all three angles.
    (EMREB)

50. Solve the simultaneous equations: $2x - 4y = 4$
    $3x + 2y = 14$. (WJEC)

51. Solve the equation, $3(x+2) = 4(2x-1)$.

52. If $v = u + ft$, give an expression for t in terms of v, u and f.

53. Write each of the following in its *simplest* form.
    (a) $a+a+a+a$ (b) $4a - a$ (c) $4a \times a$ (d) $4a \div a$
    (e) $(a+a) \times (a+a)$ (WYLEB)

54. If $x = -1$ and $y = 3$, what is the value of:
    (a) $x+y$ (b) $x-y$ (c) $y-x$ (d) xy (e) x^2 (f) $2y$?
    (WYLEB)

55. Remove the brackets and simplify:
    (a) $3(a-2b)+4(2a-b)$ (b) $(a-2)(a-3)$. (EAEB)

56. Solve the simultaneous equations: $2x - y = 1$
    $x + y = 8$
    (EAEB)

57. Use the formula $s = ut + \frac{1}{2}ft^2$ to find the value of s when $u = 8, f = 12$, and $t = \frac{1}{2}$. (EAEB)

58. If $x = 2\frac{1}{2}$, $y = -5$, $z = 3$, evaluate the following:
    (a) $3(x-z)$ (b) xyz (c) $2z^2$ (d) $\dfrac{2y}{z}$ (e) $y^2 - z^2$.

59. Simplify the following expressions:
    (a) $3(2-x)+6x$ (b) $\dfrac{x^2-y^2}{x+y}$ (c) $x(x+2)-x(x-2)$
    (d) $\dfrac{2x^3}{6xy}$.

60. Solve
   (a) $\frac{1}{4}x - 1 = 1$ (b) $4 + 3(2 - y) = 1$
   (c) $2x + y = 6$
      $3x - 2y = -1\frac{1}{2}$
   (d) if the sum of x, $2x$, $3x$ and $4x$ is 100, find the value of x.

Exercise 95

1. (a) If $a = 3$, $b = 8$, and $c = 6$, find the value of d when
   (i) $d = a(b + c)$ (ii) $ad = b - c$ (iii) $ab = c\sqrt{d}$.
   (b) Find the value of (i) $3^9 \div 3^7$ (ii) $9^{\frac{1}{2}} + 8^{\frac{1}{3}}$ (iii) $5^{-2} + 5^0$.
   (c) A farmer encloses a rectangular plot of land, x metres long and y metres wide, using a fence 100 m long. He later decides to enlarge the plot by making the new length $(x + y)$ metres, but keeping the width as y metres. The new fence is 160 m long. Form an equation in x and y for each plot of land and solve these two equations simultaneously to find the values x and y.   (*EMREB*)

2. (a) Solve the equations (i) $3x = 6$ (ii) $\frac{x}{4} = 3$.
   (b) Factorise (i) $2a + 4$ (ii) $2x + 4y + ax + 2ay$ (iii) $4x^2 - 1$.
   (c) Solve $x^2 + 5x + 1 = 0$, giving the values of x to 2 decimal places.
   (*EMREB*)

3. (a) Given that $E = mc^2$, find the value of E, if $m = 1.63 \times 10^{12}$ and $c = 3.0 \times 10^6$, leaving your answer in standard form.
   (b) If $A = (x - 5)$, $B = (x + 5)$ and $C = (2x - 5)$, find
   (i) the value of x when $A + B + C = 0$
   (ii) the values of x when $A \times B = 0$
   (iii) the values of x when $A \times B = C$, giving each value correct to 2 decimal places.   (*WYLEB*)

4. (a) (i) Write down the factors of $x^2 - y^2$.
   (ii) Hence or otherwise evaluate $(86.86)^2 - (13.14)^2$.
   (b) Find the value of D when $\frac{D}{1.3} + \frac{D}{(1.3)^2} = 23$.
   (c) Solve the equation $3x^2 - 2x - 3 = 0$, giving each root correct to 2 decimal places.   (*WYLEB*)

5. Evaluate the following, leaving your answer in standard form:
   (a) $2.3 \times 10^{-2} \times 3.0 \times 10^6$
   (b) $\frac{4.6 \times 10^{-12}}{2.3 \times 10^{-15}}$.   (*WYLEB*)

6. (a) Simplify $\frac{5x - 3y}{6} - \frac{3x - 5y}{10}$, and express your answer as a fraction in its lowest terms.

(b) Solve the simultaneous equations: $2x + 6y = 48$
$$7x - 3y = 0 \ .$$

(c) Evaluate the following: (i) $(16)^{\frac{1}{2}}$ (ii) $(4)^{-2}$ (iii) $(\frac{1}{2})^{-4}$. (*MREB*)

7. Given that $F = \dfrac{2PWk}{P+W}$, find:

(a) the value of k when $P = 6, W = 10, F = 120$
(b) a formula for P in terms of k, W and F
(c) the value of P when $k = 32, W = 5, F = 120$. (*MREB*)

8. Find the value of n in each equation:
(a) $8^n = 4^3$ (b) $8^n = 1$ (c) $8^n = \frac{1}{2}$ (d) $8^{-n} = \frac{1}{8}$. (*MREB*)

9. (a) Solve $x^2 - 5x + 2 = 0$, giving your answer correct to 2 decimal places.

(b) A cinema contained x adult customers who paid 40p each, and y child customers who paid 20p each.

(i) Write down an equation to show that the cinema contained 210 customers.

(ii) Write down a second equation to show that the receipts totalled £60.

(iii) Solve these two equations simultaneously and hence find out how many adult customers were in the cinema.

10. (a) Simplify $2y(x - y) - x(y - 2x)$.

(b) Simplify $\dfrac{1}{x-2} - \dfrac{1}{x-3}$.

(c) Solve the equation $x^2 + 5x - 1 = 0$, giving your answer correct to 2 decimal places.

(d) If $1 + R = \sqrt{\left(\dfrac{A}{P}\right)}$, find a formula for A in terms of R and P.
(*WMEB*)

11. (a) An equilateral triangle has sides of $(21 - x)$ cm, $(2x - 3y)$ cm and 6 cm. Find the value of x and y.

(b) A square has sides of $(x + 3)$ units. A rectangle has sides of $3x$ and $(x + 1)$ units.

(i) Calculate the areas of the two shapes in terms of x.

(ii) If these two areas are equal, form and solve an equation to find x.

(iii) Hence write down the dimensions of the two shapes.

12. (a) Given that $A = 6.34 \times 10^3$, $B = 3.76 \times 10^{-1}$, $C = 9.203 \times 10^2$, find the value of $\dfrac{A \times B}{C}$.

(b) Simplify $3x(2x - y) - 2y(x + 2y)$.

(c) You are given that $x = ky^2$, where k is a constant. If $x = 12$ when $y = 2$, find the value of k. Hence find the value of x when $y = 3$.

(d) Solve the quadratic equation $2x^2 + 3x - 2 = 0$. (*EMREB*)

13. On a journey, a boy walked at x miles per hour for 2 hours, and cycled at $4x$ miles per hour for 3 hours.
(a) How far did he walk in 2 hours?
(b) How far did he cycle in 3 hours?
(c) What is the total length of his journey in terms of x?
(d) If he travelled 42 miles altogether, write down an equation in x and solve it.
(e) How far did he walk?
(f) How far did he cycle?
(g) What was his average speed for the whole journey?
(Met. REB)

14. (a) If $x = -1$, find the value of: (i) $(2x+3)^2$ (ii) $2x^2 + 3$ (iii) $(2x)^2 + 3$.
(b) If $(x-4)(2x+3) = 0$, find the two values of x.
(c) If $(2x+3)^2 = 25$, find the two values of x.
(d) Solve the equation $\dfrac{5y}{4} - \dfrac{2y+3}{3} = \dfrac{y+4}{2}$. (SREB)

15. (a) If $a = 4$, find the value of:
(i) a^3 (ii) a^0 (iii) $a^{\frac{1}{4}} \times a^{\frac{1}{4}}$.
(b) If b metres of cloth can be bought for £x, how many metres of the same material can be bought for £y?
(c) Solve the simultaneous equations: $0.6a + 0.5b = 6$
$0.3a + 0.3b = 3.3$.
(d) Find the value of x if: $2^x = 4^2$. (WYLEB)

16. Find in each case the value of n:
(a) $25^n = 625$ (b) $25^n = 1$
(c) $25^n = 0.04$ (d) $25^{-n} = 0.2$ (MREB)

17. (a) Solve the equation $\dfrac{1}{x+1} = \dfrac{1}{4}$.
(b) Solve the simultaneous equations: $3x - 4y = 22$
$x + 4y = 18$.
(c) Evaluate the following: (i) $\sqrt{121}$ (ii) 2^3 (iii) $8^{\frac{1}{3}}$.
(d) (i) Express $p \times 10^n$ as one number when $p = 2.7$ and $n = 2$.
(ii) Find n when $1.4 \times 10^n = 14$. (MREB)

18. (a) A pyramid has a base 21 cm square and a height of 33 cm. Calculate its volume.
(b) A rectangular box stands on a base 21 cm square. Its volume is 4851 cm^3. Find its height.
(c) Find the radius of a sphere whose volume is 4851 cm^3. (Take π as $\frac{22}{7}$) (WMEB)

19. (a) Factorise (i) $4(x-6) - y(x-6)$ (ii) $3x^2 - 5x + 2$.

(b) Find the value of (i) $9^{\frac{1}{2}}$ (ii) $16^{-\frac{1}{2}}$ (iii) $16^{\frac{3}{4}}$.

(c) In the figure, OAB and OCD are straight lines.

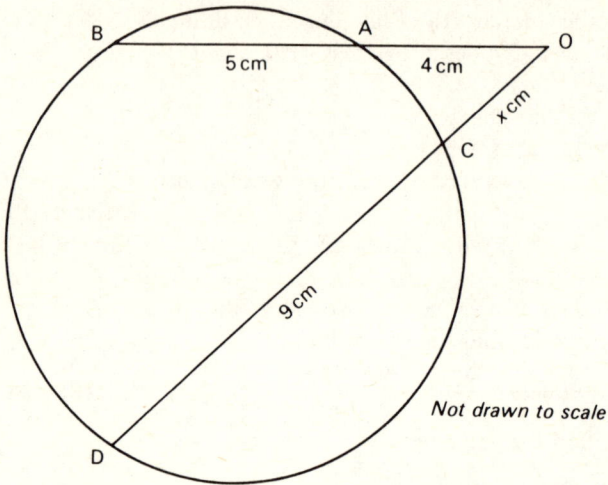

Not drawn to scale

AB = 5 cm, OA = 4 cm, DC = 9 cm, OC = x cm.

Form an equation for x and solve it to find the length of OC.

(WMEB)

20. The diagram shows a water tank which is made up of an open cylinder and a cone. The diameter of the cylinder is 10 m and its height is 6 m. The vertical height of the cone is 2 m. Calculate, correct to 2 decimal places
(a) the volume of the tank,
(b) its total surface area.

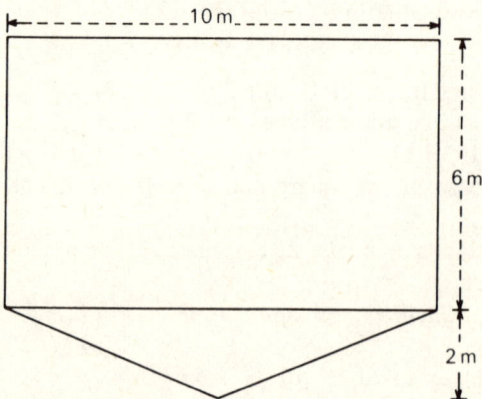

Not drawn to scale

(Volume of a cylinder $= \pi r^2 h$, volume of cone $= \frac{1}{3}\pi r^2 h$, surface area of an open cylinder $= 2\pi rh$, curved surface area of cone $= \pi r\sqrt{(r^2 + h^2)}$; take $\pi = 3.14$)

21. (a) The figure shows a rectangle ABCD, in which EF is parallel to DC, AB = BF = y, and AD = x.

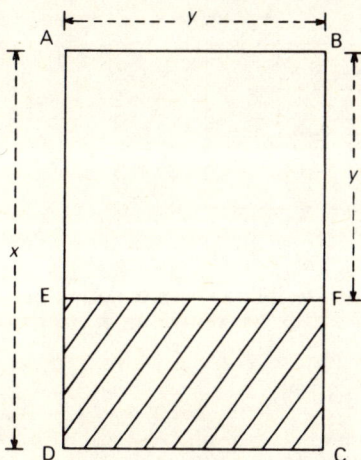

(i) Express the area of the shaded part of the figure in terms of x and y.

(ii) If $x:y = 3:2$ and the perimeter ABCD is 30 cm, find the values of x and y.

(iii) Using these values of x and y, find what fraction of the whole is the unshaded part.

(b) The cost ($£c$) of hiring a motor coach is found by adding the driver's wage ($£w$) to a mileage charge of p pence per mile. If the coach is hired for a journey of m miles,

(i) express the charge c in terms of w, p and m

(ii) find p in terms of c, w and m.   (WJEC)

22. (a) Solve the simultaneous equations: $x + 4y = 12$
                                               $3x + 10y = 29.$

(b) Find the two values of x which satisfy the equation $x^2 = 4x$.

(c) The diagram on page 90, which is not drawn to scale, shows a lawn surrounded by a path which is 2 m wide. The whole garden is x m long and 10 m wide. Write down:

(i) the width of the lawn,

(ii) the length of the lawn,

(iii) the area of the lawn.

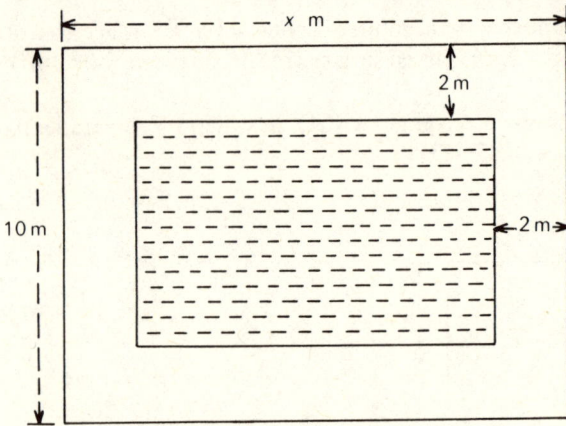

If the total area of the path is 80 m², write down an equation and hence find the value of x.   (*EMREB*)

Answers

Exercise 1
(1) 11   (2) 16   (3) 17   (4) 20   (5) 23   (6) 21   (7) 2   (8) 2   (9) 7   (10) 6
(11) 2   (12) 3   (13) -3 (14) -5 (15) -2 (16) -11 (17) -3 (18) -8
(19) 1   (20) 5   (21) 1   (22) 14   (23) 12   (24) 2   (25) -5 (26) -2
(27) -2 (28) -5 (29) -15 (30) -6 (31) -17 (32) -17 (33) -12
(34) -3 (35) -23 (36) -13

Exercise 2
(1) 6   (2) -6 (3) -12 (4) 5   (5) 0   (6) -8 (7) 11   (8) -8 (9) -11
(10) -4 (11) 8   (12) -5 (13) 5   (14) -7 (15) -21 (16) -9 (17) 1
(18) 4   (19) -1 (20) 0   (21) 3   (22) -3 (23) -3 (24) -20 (25) 2
(26) 38   (27) -15 (28) -5 (29) 6   (30) 2

Exercise 3
(1) 24   (2) 63   (3) 66   (4) 25   (5) 36   (6) 72   (7) -24 (8) -63
(9) -66 (10) -25 (11) -36 (12) -72 (13) -24 (14) -63 (15) -66
(16) -25 (17) -36 (18) -72 (19) 24   (20) 63   (21) 66   (22) 25
(23) 36   (24) 72

Exercise 4
(1) -6 (2) 32   (3) 30   (4) -72 (5) 16   (6) 16   (7) -16 (8) -33
(9) -33 (10) -56 (11) 0   (12) 25   (13) 25   (14) 0   (15) 54   (16) 49
(17) -2 (18) -12 (19) 54   (20) 54   (21) -1 (22) 1   (23) 12   (24) -12
(25) -21 (26) 64   (27) 0   (28) -36 (29) 17   (30) -14

Exercise 5
(1) 3   (2) $\frac{1}{2}$ (3) 1   (4) 9   (5) $\frac{1}{3}$ (6) 6   (7) -3 (8) $-\frac{1}{2}$ (9) -1
(10) -9 (11) $-\frac{1}{3}$ (12) -6 (13) -3 (14) $-\frac{1}{2}$ (15) -1 (16) -9
(17) $-\frac{1}{3}$ (18) -6 (19) 3   (20) $\frac{1}{2}$ (21) 1   (22) 9   (23) $\frac{1}{3}$ (24) 6

Exercise 6
(1) -5 (2) 10   (3) 8   (4) -3 (5) $\frac{1}{2}$ (6) $\frac{1}{2}$ (7) -2 (8) $-2\frac{1}{3}$ (9) -4
(10) -4 (11) 1   (12) -4 (13) -4 (14) 0   (15) -3 (16) 5   (17) $\frac{1}{2}$
(18) $-4\frac{1}{2}$

Exercise 7
(1) 10   (2) 8   (3) 21   (4) 11   (5) 14   (6) 9   (7) 2   (8) 2   (9) 7   (10) 1
(11) 0   (12) 1   (13) -2 (14) -5 (15) -11 (16) 0   (17) -3 (18) -2
(19) 2   (20) 4   (21) 8   (22) 5   (23) 2   (24) 1   (25) -2 (26) -7 (27) -1

(28) -3 (29) -6 (30) -9 (31) -10 (32) -19 (33) -9 (34) -30
(35) -11 (36) -4

Exercise 8
(1) 2   (2) 2   (3) -3 (4) -11 (5) 0   (6) 7   (7) 10   (8) 8   (9) 9
(10) 7   (11) 14   (12) 11   (13) -14 (14) -10 (15) -12 (16) -25
(17) -10 (18) -9 (19) -2 (20) 5 (21) -7 (22) 0 (23) 3 (24) 2

Exercise 9
(1) -1 (2) 4   (3) -80 (4) 0   (5) 39   (6) -5 (7) 1   (8) 2   (9) -5
(10) -7 (11) -4 (12) 14   (13) 1   (14) 8   (15) -9 (16) 0   (17) 21
(18) -21 (19) 10   (20) 0

Exercise 10
(1) $4a$ (2) $-2a$ (3) $5b$ (4) $2d$ (5) $-5h$ (6) $-6x$ (7) $15c$
(8) $-8e+8f$ (9) $-4a$ (10) $-15y$ (11) $5p$ (12) 0   (13) $2a+2b$
(14) $2a-2b$ (15) $-2a+2b$ (16) $c+2d$ (17) $2x+12y$ (18) $p-2r$
(19) $4a-3b+2c$ (20) $3x-10y$ (21) $15r+8s$ (22) $-3a-5b$ (23) $8f-4g$
(24) $4s-11t-r$

Exercise 11
(1) $5x+2x^2-x^3$ (2) $2x-5x^2+x^4$ (3) $-3y+y^2+2y^3$ (4) $7-x+4x^2$
(5) $-2+4x-3x^2+3x^3$ (6) $8+t^3$ (7) $-2+9m-m^5$ (8) $6x-5x^2-x^3$
(9) $-3a^3+6a^2+10a$ (10) $3x^3-2x^2+7x$ (11) a^3-2a^2+5a (12) $4t$ (13) 7
(14) $8x^5-x^2+7x$ (15) y^3-3y^2-y (16) $-4a^3+a^2-a$

Exercise 12
(1) $2g$ (2) $4t$ (3) abc (4) $2xy$ (5) $12rs$ (6) $18hk$ (7) $4ab$ (8) $8xyz$
(9) $4ts$ (10) $6pqrs$ (11) $6abc$ (12) $5pqr$ (13) $12hkl$ (14) $6mnpq$ (15) $72r$
(16) $8abc$ (17) $36lmnp$ (18) $35rs$ (19) $8abcd$ (20) $18efgh$ (21) a^3 (22) $8x^2$
(23) $9y^2$ (24) $8t^3$ (25) $15b^2$ (26) $6x^3$ (27) y^5 (28) $12y^2$ (29) $6p^2$
(30) $8x^3$ (31) $6x^2$ (32) $8x^3$

Exercise 13
(1) x^5 (2) y^6 (3) a^4 (4) b^6 (5) x^7 (6) y^{10} (7) r^3s^3 (8) x^3y^2 (9) x^5y
(10) y^5z^2 (11) p^2r (12) x^4y (13) a^5b (14) x^3y^5 (15) pr^5 (16) c^4d^2
(17) c^2d^4 (18) x^2y^2 (19) x^3y^4 (20) p^3r^4 (21) m^4n^7 (22) m^4n^4 (23) y^2z^2
(24) b^5c^4

Exercise 14
(1) $6x^5$ (2) $8y^6$ (3) $20a^4$ (4) $36b^6$ (5) $10x^7$ (6) $8y^{10}$ (7) $6x^2y^2z$
(8) $12r^2t^3$ (9) $2xy^3z$ (10) $12xy^2z$ (11) $6a^3b^4c$ (12) x^6

Exercise 15
(1) x^3 (2) y^6 (3) a (4) y (5) 1   (6) x^4 (7) $\dfrac{1}{x^3}$ (8) $\dfrac{1}{y^6}$ (9) $\dfrac{1}{a}$ (10) $\dfrac{1}{y}$

(11) 1   (12) $\dfrac{1}{x^4}$ (13) x^2y^2 (14) ab^2 (15) at (16) x^2yz^2 (17) s (18) bc

(19) $\dfrac{1}{x^2y}$ (20) $\dfrac{1}{pqr}$ (21) $\dfrac{1}{st^2}$ (22) $\dfrac{1}{xy^2}$ (23) 1   (24) $\dfrac{1}{p^2}$ (25) $\dfrac{x^5}{y^2}$ (26) $\dfrac{y}{z}$

(27) $\dfrac{t^2}{2s}$ (28) $\dfrac{5t}{s}$ (29) $\dfrac{km}{2}$ (30) $\dfrac{x}{y}$

Exercise 16

(1) $2x^2$ (2) $\dfrac{x^5}{5}$ (3) 3   (4) $6a$ (5) $\dfrac{6}{a^2}$ (6) $\dfrac{2x}{7}$ (7) $\dfrac{x}{2y^2}$ (8) $\dfrac{5c^2}{a}$ (9) $\dfrac{9x}{2y^2}$

(10) $\dfrac{a}{4b}$ (11) $\dfrac{2a}{3}$ (12) $\dfrac{xy^2}{5}$ (13) $\dfrac{a}{b}$ (14) $\dfrac{3z}{xy}$ (15) $\dfrac{m}{n}$ (16) $\dfrac{3ac}{2b}$ (17) 1

(18) $\dfrac{3x}{4z}$ (19) $\dfrac{3a}{bc}$ (20) $\dfrac{5}{6xy^2}$

Exercise 17a

(1) 12   (2) 7   (3) 8   (4) 26   (5) -12 (6) 36   (7) 14   (8) 19   (9) 29
(10) 16   (11) 21   (12) -84

Exercise 17b

(1) -6 (2) 2   (3) 0   (4) 12   (5) 6   (6) -21 (7) -9 (8) 9   (9) 4
(10) -7 (11) 4   (12) -3

Exercise 18a

(1) 5   (2) 1   (3) -1 (4) 6   (5) 6   (6) 8   (7) 0   (8) 15   (9) 10   (10) 14
(11) 4   (12) 1

Exercise 18b

(1) 3   (2) -7 (3) 7   (4) -10 (5) 8   (6) -9 (7) -12 (8) 9   (9) 6
(10) 11   (11) 5   (12) 0

Exercise 19a

(1) 6   (2) 0   (3) 45   (4) 6   (5) 0   (6) 81   (7) 12   (8) 18   (9) 0   (10) 36
(11) 8   (12) 16

Exercise 19b

(1) -2 (2) -5 (3) -5 (4) -3 (5) 0   (6) 16   (7) 8   (8) 15   (9) 0
(10) 8   (11) 4   (12) 0

Exercise 20a

(1) 5   (2) 3   (3) 2   (4) $\frac{1}{2}$ (5) 4   (6) 1   (7) $\frac{1}{3}$ (8) 1   (9) $\frac{1}{4}$ (10) $1\frac{1}{2}$
(11) $1\frac{1}{2}$ (12) 1

Exercise 20b

(1) 5   (2) -2 (3) -3 (4) $1\frac{1}{2}$ (5) $\frac{2}{3}$ (6) -3 (7) 1   (8) 2   (9) -1
(10) $\frac{1}{5}$ (11) -3 (12) -2

Exercise 21

$x+y$	$x-y$	xy	$\dfrac{x}{y}$
8	2	15	$1\frac{2}{3}$
4	-4	0	0
2	0	1	1
-1	-7	-12	$-1\frac{1}{3}$
1	3	-2	-2
-5	1	6	$\frac{2}{3}$

Exercise 22
(1) 2   (2) -4 (3) -10 (4) 11   (5) -1 (6) 8   (7) $3\frac{1}{2}$ (8) -7 (9) -13
(10) 0   (11) 12   (12) 0   (13) -2 (14) -10 (15) 30   (16) -1 (17) 2
(18) 0   (19) 6   (20) 2   (21) 56   (22) 12   (23) 0   (24) 12   (25) 32   (26) -4
(27) 0   (28) 5   (29) -3 (30) 0   (31) 21   (32) 12

Exercise 23
(1) 20   (2) 12   (3) $5x+2$ (4) $9+2x$ (5) $6x+3$ (6) $9y$ (7) 14   (8) 2
(9) $5x-2$ (10) $9-2x$ (11) $6x-3$ (12) $3y$ (13) 2   (14) -6 (15) $-x-2$
(16) $-1-2x$ (17) $-2x+3$ (18) $5y$ (19) 8   (20) 4   (21) $-x+2$
(22) $-1+2x$ (23) $-2x-3$ (24) $-y$

Exercise 24
(1) 17   (2) 12   (3) $3a-b$ (4) $2c$ (5) $a-2b+c$ (6) $4a$ (7) $2y$
(8) $8a-b$ (9) $x-y-z$ (10) $y-4x$ (11) a (12) $a-b$ (13) 10   (14) $x+9$
(15) $6a-1$ (16) $9-x$ (17) $2a-2c$ (18) $2p+3q-4r$

Exercise 25a
(1) $3x+3y$ (2) $7s+7t$ (3) $4m+24$ (4) $6+2k$ (5) $6a+15b$ (6) $3x+5$
(7) $5a+10b+30$ (8) $6x+6y+1\frac{1}{2}$ (9) $2a-2b$ (10) $3x-3y$ (11) $5t-10$
(12) $12-3d$ (13) $3x-y$ (14) $14a-21b$ (15) $4x-8y-12$ (16) $12a-6b+12$
(17) $-3x-3y$ (18) $-7s-7t$ (19) $-4m-24$ (20) $-6-2k$ (21) $-6a-15b$
(22) $-2x-4\frac{1}{2}$ (23) $-5a-10b-30$ (24) $-6x-6y-4\frac{1}{2}$ (25) $-2a+2b$
(26) $-3x+3y$ (27) $-5t+10$ (28) $-12+3d$ (29) $-2x+\frac{1}{2}y$ (30) $-14a+21b$
(31) $-4x+8y+12$ (32) $-12a+6b-12$

Exercise 25b
(1) $ax+ay$ (2) s^2+st (3) $km+6k$ (4) $3k+k^2$ (5) $2ax+5bx$ (6) $ab-b^2$
(7) $ax-ay$ (8) t^2-2t (9) $4e-de$ (10) $12x^2-4xy$ (11) $-ax-ay$
(12) $-s^2-st$ (13) $-km-6k$ (14) $-3k-k^2$ (15) $-2ax-5bx$ (16) $-ab+b^2$
(17) $-ax+ay$ (18) $-t^2+2t$ (19) $-4e+de$ (20) $-12x^2+4xy$

Exercise 26
(1) $6a^2 - 8ab$ (2) $-xy + x^2$ (3) $2x^2y + 2xy^2$ (4) $3x^2 - x^3$ (5) $-6y - 2xy + 2y^2$
(6) $2a^3 + 2a^2 + a$ (7) $3x - 2y + 8$ (8) $x^2yz + xy^2z + xyz^2$ (9) $-4x^2 + 8xy - 12xz$
(10) $4ab - 4ac - 2a$ (11) $x - tx$ (12) $p^3q - pq^3$

Exercise 27
(1) $5x + 16$ (2) $11x + 1$ (3) $2x + 34$ (4) $16x + 7$ (5) $5x + 1$ (6) $x - 21$
(7) $3x + 7$ (8) $3x - 7$ (9) $18 - 6x$ (10) $x + 1$ (11) $14x + 6$ (12) $5 - 7x$

Exercise 28
(1) $am + an + bm + bn$ (2) $cx + cy + dx + dy$ (3) $a^2 + 5a + 6$ (4) $b^2 + 5b + 4$
(5) $x^2 + 6x + 8$ (6) $12 + 7a + a^2$ (7) $ax + 2a + bx + 2b$ (8) $2c + 2d + cy + dy$
(9) $a^2 + 10a + 16$ (10) $x^2 + 10x + 25$ (11) $xy + ax + ay + a^2$ (12) $t^2 + 15t + 56$
(13) $3y + 3z + xz + xy$ (14) $25 + 10p + p^2$ (15) $x^2 + 3x + 2$ (16) $a^2 + 9a + 18$
(17) $y^2 + 9y + 20$ (18) $6 + 5y + y^2$ (19) $x^2 + 12x + 32$ (20) $x^2 + 9x + 14$

Exercise 29
(1) $ax - ay + bx - by$ (2) $x^2 - x - 12$ (3) $x^2 + x - 12$ (4) $6 + x - x^2$ (5) $a^2 - a - 30$
(6) $x^2 + 4x - 5$ (7) $x^2 - 9$ (8) $a^2 - 6a - 16$ (9) $t^2 - t - 2$ (10) $12 - y - y^2$
(11) $x^2 - 2x - 35$ (12) $x^2 + 4x - 12$ (13) $x^2 + 2x - 35$ (14) $x^2 - 25$ (15) $x^2 + 7x - 30$
(16) $x^2 - y^2$

Exercise 30
(1) $ax + ay - bx - by$ (2) $x^2 + x - 12$ (3) $x^2 - x - 12$ (4) $4 - 3x - x^2$
(5) $a^2 + 2a - 15$ (6) $x^2 - 3x - 18$ (7) $x^2 - 4$ (8) $ab + 7a - 5b - 35$ (9) $t^2 + t - 6$
(10) $8 + 2y - y^2$ (11) $x^2 + 4x - 60$ (12) $x^2 - 2x - 3$ (13) $x^2 - x - 2$ (14) $x^2 - 36$
(15) $x^2 - 10x - 24$ (16) $x^2 + 2x - 35$

Exercise 31
(1) $ax - bx - ay + by$ (2) $x^2 - 9x + 20$ (3) $x^2 - 4x + 3$ (4) $a^2 - 5a + 6$
(5) $12 - 7y + y^2$ (6) $a^2 - 12a + 36$ (7) $6t - 8 - t^2$ (8) $x^2 - 3x + 2$
(9) $x^2 - 10x + 21$ (10) $2ax - a^2 - x^2$ (11) $y^2 - 11y + 30$ (12) $x^2 - 2x + 1$
(13) $x^2 - 12x + 20$ (14) $x^2 - 7x + 12$ (15) $x^2 - 16x + 63$ (16) $x^2 - 2xy + y^2$

Exercise 32
(1) $x^2 + 8x + 16$ (2) $x^2 + 2x + 1$ (3) $x^2 + 12x + 36$ (4) $x^2 - 4x + 4$
(5) $x^2 - 10x + 25$ (6) $x^2 - 2ax + a^2$ (7) $x^2 + 16x + 64$ (8) $x^2 - 14x + 49$
(9) $a^2 + 2ax + x^2$ (10) $16 - 8t + t^2$ (11) $x^2 + 20x + 100$ (12) $x^2 - 18x + 81$

Exercise 33
(1) $x^2 - y^2$ (2) $x^2 - 16$ (3) $x^2 - 1$ (4) $m^2 - n^2$ (5) $x^2 - 100$ (6) $49 - x^2$
(7) $x^2 - 4$ (8) $a^2 - x^2$ (9) $x^2 - 36$ (10) $16 - x^2$ (11) $x^2 - 49$ (12) $1 - x^2$

Exercise 34
(1) $4x^2 + 11xy + 6y^2$ (2) $12x^2 + 12x + 3$ (3) $4x^2 + 20x + 25$ (4) $10x^2 + 11x - 6$
(5) $6x^2 - 11x - 7$ (6) $4x^2 - 31x - 45$ (7) $12 - 7x - 10x^2$ (8) $2y^2 + 5xy - 3x^2$
(9) $10y^2 + y - 3$ (10) $9x^2 - 24xy + 16y^2$ (11) $18x^2 - 15x + 2$ (12) $8xy - 3y^2 - 4x^2$

Exercise 35
(1) $a^2 + 2a - 48$ (2) $2x^2 + 11x + 12$ (3) $x^2 - y^2$ (4) $12x^2 - 11x + 2$
(5) $10 - 3y - y^2$ (6) $36x^2 + 12x + 1$ (7) $25 - 10x + x^2$ (8) $p^2 - 3p + 2$

(9) $a^2 - 3a - 18$ (10) $6x^2 + 11x + 4$ (11) $m^2 + 24m + 144$ (12) $4x^2 + 12x + 9$

(13) $6x^2 + 13xy + 6y^2$ (14) $100x^2 - 100$ (15) $12x^2 - 12x - 72$ (16) $a^2 - 5ab + 6b^2$

Exercise 36

(1) $2(x + 4y)$ (2) $3(2r + s + 3t)$ (3) $4(5x - 2y)$ (4) $x(x + y)$ (5) $c(4b - 3d)$

(6) $5x(x - 2y)$ (7) $xy(1 + x)$ (8) $5b(a - b + 3c)$ (9) $-4(3a + b)$

(10) $-2(x - 5y + 2z)$ (11) $2x(2x - y)$ (12) $4x(y + 3z)$ (13) $9(d - 5e)$

(14) $21a(b + 3)$ (15) $xy(x + y + 1)$

Exercise 37

(1) $5(a - 2b)$ (2) $3(2x + y - 4z)$ (3) $7(m + 8n)$ (4) $2(7r - 4s - 6t)$

(5) $3(x + 2y + 2z)$ (6) $8(3x - 4y + 2z)$ (7) $a(x + y)$ (8) $z(c - d)$ (9) $x(x + y)$

(10) $p(7q - 9r)$ (11) $y(3x + 2y - 7)$ (12) $b(ac - 2c + 4b)$ (13) $3b(a + 3c)$

(14) $2x(2x - 3y)$ (15) $7d(4c + d)$ (16) $3xy(xy - 2z)$ (17) $xyz(x + y - z)$

(18) $5b(c - 5ab)$ (19) $4x(3y + 2x)$ (20) $x^2(x^2 - 1)$

Exercise 38

(1) $(3 + a)(x - y)$ (2) $(x + y)(a + b)$ (3) $(x + y)(x - 2)$ (4) $(a - b)(2x + 1)$

(5) $(2x - 5y)(a - 5)$ (6) No factors   (7) $(4 + y)(x^2 - 1)$ (8) $(a - d)(b + c)$

(9) $(x - y)(3 - b)$ (10) $(2x + y)(x + 1)$

Exercise 39

(1) $(3 + a)(x + y)$ (2) $(b + 5)(a + 4)$ (3) $(a + 2)(x + y)$ (4) $(a + b)(a + 2)$

(5) $(x + 3)(t + 3)$ (6) $(5 - b)(x + y)$ (7) $(x - 6)(a + 3)$ (8) $(4 - t)(x + y)$

(9) $(m - n)(m + 2)$ (10) $(y - x)(2 + x)$ (11) $(2 + a)(x - y)$ (12) $(y + 5)(x - 4)$

(13) $(7 + x)(x - y)$ (14) $(4 + y)(4 - x)$ (15) $(x + 3)(a - 3)$ (16) $(4 - a)(x - y)$

(17) $(a - 2x)(3x - 1)$ (18) $(4 - t)(x - y)$ (19) $(a - 1)(x - y)$ (20) $(y - x)(5 - x)$

Exercise 40

(1) $(3 + x)(a + 2)$ (2) $(x + y)(y + 2)$ (3) $(a - d)(b + c)$ (4) $(a - 7)(x - 2)$

(5) No factors   (6) $(x - 3)(x + y)$ (7) $(3 + 2y)(2 + x)$ (8) $(y - x)(2 - a)$

Exercise 41

(1) $(x + 4)(x + 1)$ (2) $(x + 4)(x + 3)$ (3) $(x + 3)(x + 1)$ (4) $(x + 1)(x + 8)$

(5) $(x + 1)(x + 6)$ (6) $(x + 2)(x + 7)$ (7) $(y + 1)(y + 5)$ (8) $(a + 2)(a + 16)$

(9) $(x + 2)(x + 11)$ (10) $(t + 6)(t + 2)$ (11) $(x + 1)(x + 1)$ (12) $(x + 4)(x + 9)$

Exercise 42

(1) $(x - 2)(x - 3)$ (2) $(x - 1)(x - 5)$ (3) $(x - 2)(x - 5)$ (4) $(x - 6)(x - 4)$

(5) $(a - 5)(a - 5)$ (6) $(y - 2)(y - 2)$ (7) $(x - 7)(x - 1)$ (8) $(x - 6)(x - 2)$

(9) $(x - 6)(x - 5)$ (10) $(x - 2)(x - 18)$ (11) $(t - 4)(t - 25)$ (12) $(x - 4)(x - 5)$

Exercise 43

(1) $(x + 3)(x - 2)$ (2) $(x - 3)(x + 2)$ (3) $(x + 10)(x - 1)$ (4) $(x + 7)(x - 1)$

(5) $(x + 3)(x - 1)$ (6) $(x - 4)(x + 3)$ (7) $(t - 9)(t + 1)$ (8) $(x + 8)(x - 3)$

(9) $(y + 4)(y - 6)$ (10) $(x + 2)(x - 1)$ (11) $(x - 6)(x + 5)$ (12) $(x + 6)(x - 3)$

Exercise 44

(1) $(x + 3)^2$ (2) $(x - 10)^2$ (3) $(x + y)^2$ (4) $(a - 6)^2$ (5) $(y - 1)^2$ (6) $(2 + x)^2$

(7) $(x + 7)^2$ (8) $(x - 7)^2$ (9) $(t + a)^2$ (10) $(x + 5)^2$ (11) $(x - 9)^2$ (12) $(2x + 3)^2$

Exercise 45

(1) $(x-2)(x+2)$ (2) $(x-9)(x+9)$ (3) $(x-8)(x+8)$ (4) $(x-7)(x+7)$
(5) $(x-12)(x+12)$ (6) $(x-1)(x+1)$ (7) $(x-y)(x+y)$ (8) $(x-3)(x+3)$
(9) $(4-x)(4+x)$ (10) $(x-2y)(x+2y)$ (11) $(3x-5)(3x+5)$ (12) $(2p-3q)(2p+3q)$
(13) $(x-13)(x+13)$ (14) $(xy-2)(xy+2)$ (15) $(15-2x)(15+2x)$

Exercise 46

(1) $(2x+1)(x+2)$ (2) $(3x-1)(x-4)$ (3) $(2x-3)(x+1)$ (4) $(2x+5)(x-1)$
(5) $(5x-2)(x-3)$ (6) $(3x+2)(x+3)$ (7) $(3x-4)(x-1)$ (8) $(1-5x)(5+x)$

Exercise 47

(1) $(a+9)(a-1)$ (2) $(x+4)^2$ (3) $(5-x)(5+x)$ (4) $(3+a)(x+y)$
(5) $(x-4)(x-3)$ (6) $(y-11)(y+1)$ (7) $(2a-b)(2a+b)$ (8) $(x-1)^2$
(9) $(t+x)(t-1)$ (10) $(x+8)(x+5)$ (11) $(6+x)(2+x)$ (12) $(12-y)(2-y)$
(13) $(20-t)(20+t)$ (14) $(x-5)(x+2)$ (15) $(x-8)^2$ (16) $(3+t)^2$
(17) $(2-y)(2+x)$ (18) $(x-11)(x+11)$ (19) $(2x+3)(x+2)$ (20) $(2x-2)(x+4)$

Exercise 48

(1) 12   (2) 4   (3) 15   (4) $16ab$ (5) $5b^2$ (6) $15az$ (7) $3t^2$ (8) $15a^2b$
(9) 4   (10) 1   (11) 9   (12) a (13) a (14) 3   (15) x (16) 1   (17) 2
(18) $9abc$ (19) 5   (20) $2at$ (21) $15mp$ (22) x^4 (23) $16xy$ (24) ab

Exercise 49

(1) $\dfrac{a}{2b}$ (2) $\dfrac{x}{y}$ (3) $\frac{2}{3}$ (4) $\dfrac{x}{y}$ (5) $\dfrac{x}{5t}$ (6) $\dfrac{1}{3h}$ (7) $\frac{1}{2}$ (8) 1   (9) $\dfrac{2x}{3y}$ (10) $\dfrac{1}{3c}$

(11) $\dfrac{n}{3}$ (12) $\frac{1}{5}$ (13) $\dfrac{1}{3x^2}$ (14) $\dfrac{2z}{5x}$ (15) xy

Exercise 50

(1) $\frac{10}{4}$ (2) $\dfrac{7x}{3}$ (3) $\dfrac{7}{a}$ (4) $\frac{4}{2}$ (5) $\dfrac{3x}{7}$ (6) $\dfrac{6}{x}$ (7) $\dfrac{(b-c)}{x}$ (8) $\dfrac{9}{xy}$ (9) $\dfrac{6a}{bc}$

(10) $\dfrac{(a+b)}{bc}$ (11) $\dfrac{(2x-2a)}{8}$ (12) $\dfrac{(3x+5y)}{ab}$ (13) $\dfrac{8x}{3}$ (14) 0   (15) $\dfrac{7ab}{x}$

(16) $\dfrac{(x-y)}{a^2}$ (17) $\dfrac{9}{abc}$ (18) $\dfrac{(2m-3n)}{4t}$ (19) $\dfrac{9}{8x}$ (20) $\dfrac{(a+3b)}{x}$

Exercise 51

(1) $\dfrac{x}{6}$ (2) $\dfrac{9x}{6}$ (3) $\dfrac{x}{20}$ (4) $\dfrac{23x}{12}$ (5) $\dfrac{9x}{40}$ (6) $\dfrac{14xy}{15}$ (7) $\dfrac{7x}{24}$ (8) $\dfrac{16x}{21}$ (9) $\dfrac{5x}{18}$

(10) $\dfrac{(3y+4x)}{xy}$ (11) $\dfrac{(2bx-ay)}{ab}$ (12) $\dfrac{(a^2+b^2)}{ab}$ (13) $\dfrac{(3+4x)}{x^2}$ (14) $\dfrac{(20+3x)}{4x}$

(15) $\dfrac{(6z-2x)}{xyz}$ (16) $\dfrac{(5-2a)}{a}$ (17) $\dfrac{(7a^2-7b^2)}{ab}$ (18) $\dfrac{(5x+2y)}{5}$

Exercise 52

(1) $\dfrac{(3x+5)}{3}$ (2) $\dfrac{(6x+3)}{4}$ (3) $\dfrac{(x+1)}{5}$ (4) $\dfrac{(7x+11)}{2}$ (5) $\dfrac{(3x-16)}{10}$

A.—G

(6) $\dfrac{(-2x+4)}{6}$ (7) $\dfrac{(7x+8)}{6}$ (8) $\dfrac{(30x-11)}{12}$ (9) $\dfrac{(x+1)}{4}$ (10) $\dfrac{(7x+30)}{15}$

(11) $\dfrac{(14x-27)}{12}$ (12) $\dfrac{(11x-4)}{20}$ (13) $\dfrac{(22x+3)}{4}$ (14) $\dfrac{(7x-10)}{18}$ (15) $\dfrac{(x+17)}{6}$

(16) $\dfrac{(2x+11)}{12}$ (17) $\dfrac{(20x+25)}{12}$ (18) $\dfrac{(2x+6)}{5}$

Exercise 53

(1) $\dfrac{(7x+5)}{6}$ (2) $\dfrac{(10x-53)}{12}$ (3) $\dfrac{(46x-29)}{20}$ (4) $\dfrac{(3x+6)}{21}$ (5) $\dfrac{(26x+27)}{30}$

(6) $\dfrac{(27x-24)}{10}$ (7) $\dfrac{2x}{(x-1)(x+1)}$ (8) $\dfrac{(7x+3)}{(2x-3)(x+3)}$ (9) $\dfrac{10}{(2x-5)(2x+5)}$

(10) $\dfrac{(10x-6)}{(x-5)(2x+1)}$ (11) $\dfrac{(x^2+xy+y^2)}{(x-y)(x+2y)}$ (12) $\dfrac{(5y-7x)}{2x(x+y)}$

Exercise 54

(1) $\dfrac{2x}{y}$ (2) $\dfrac{xy}{5}$ (3) $\dfrac{2x}{3y}$ (4) $\dfrac{ay}{2}$ (5) $\frac{1}{4}$ (6) $\dfrac{a}{bx}$ (7) $\dfrac{3xy}{4}$ (8) $\dfrac{a}{d}$ (9) $\dfrac{2y}{3}$

(10) $\dfrac{2a}{b}$ (11) $\dfrac{xy}{12z^2}$ (12) $\dfrac{3a^2}{c}$ (13) $\dfrac{1}{abxy}$ (14) $\dfrac{4x^2}{11y^2}$ (15) $\dfrac{4x}{3}$ (16) $\dfrac{3a}{b}$

Exercise 55

(1) $\dfrac{x}{2y}$ (2) $\dfrac{a^2}{2c}$ (3) $\dfrac{3}{2y}$ (4) $\dfrac{z}{y}$ (5) $\dfrac{5x^4}{3y^4}$ (6) $\frac{5}{3}$ (7) $\dfrac{3x}{2y}$ (8) $\dfrac{1}{(20xy)}$ (9) $\dfrac{3y}{8x}$

(10) y

Exercise 56

(1) 3 (2) 8 (3) 7 (4) 8 (5) 1 (6) 5 (7) $2\frac{1}{2}$ (8) $1\frac{1}{2}$ (9) 0 (10) -5
(11) 8 (12) 14 (13) 5 (14) 20 (15) 6 (16) $5\frac{1}{2}$ (17) $26\frac{1}{2}$ (18) 8 (19) 1
(20) 2 (21) 3 (22) 10 (23) 4 (24) 4 (25) $4\frac{1}{2}$ (26) $3\frac{1}{3}$ (27) 8 (28) $\frac{1}{2}$
(29) $\frac{3}{5}$ (30) $\frac{7}{3}$ (31) 20 (32) 9 (33) 6 (34) 21 (35) 7 (36) 50 (37) $7\frac{1}{2}$
(38) 48 (39) 0 (40) 32

Exercise 57

(1) 6 (2) $5\frac{1}{2}$ (3) 3 (4) $1\frac{2}{3}$ (5) -11 (6) 3 (7) -2 (8) 1 (9) 6 (10) 0
(11) 2 (12) 1 (13) 6 (14) 7 (15) $3\frac{2}{5}$ (16) 6 (17) 0 (18) $3\frac{1}{5}$ (19) 10
(20) 1 (21) 3 (22) 5 (23) -9 (24) $-\frac{2}{3}$

Exercise 58

(1) 1 (2) -2 (3) -2 (4) 3 (5) $-3\frac{1}{2}$ (6) 4 (7) 30 (8) 1 (9) 1
(10) 8 (11) 3 (12) 5 (13) 2 (14) -4 (15) -5 (16) 3

Exercise 59

(1) 12 (2) 9 (3) 12 (4) 5 (5) 12 (6) 6 (7) -21 (8) 15 (9) 20
(10) $12\frac{1}{2}$ (11) 4 (12) 9

Exercise 60
(1) 6   (2) $\frac{6}{7}$ (3) 20   (4) 12   (5) 8   (6) 3   (7) -15 (8) 10   (9) $6\frac{2}{3}$

Exercise 61
(1) 1   (2) -2 (3) 3   (4) -9 (5) 2   (6) $12\frac{1}{2}$ (7) 11   (8) 25   (9) 2
(10) -1 (11) -3 (12) 12

Exercise 62
(1) 5   (2) -7 (3) -9 (4) 20   (5) $2\frac{1}{5}$ (6) $-\frac{8}{9}$ (7) 6   (8) 5   (9) -4
(10) 3

Exercise 63
(1) $\left.\begin{array}{l} x=6 \\ y=1 \end{array}\right\}$ (2) $\left.\begin{array}{l} x=3 \\ y=5 \end{array}\right\}$ (3) $\left.\begin{array}{l} x=4 \\ y=-2 \end{array}\right\}$ (4) $\left.\begin{array}{l} x=\frac{1}{2} \\ y=-1 \end{array}\right\}$ (5) $\left.\begin{array}{l} x=-3 \\ y=8 \end{array}\right\}$ (6) $\left.\begin{array}{l} x=-3 \\ y=-4 \end{array}\right\}$

(7) $\left.\begin{array}{l} x=2 \\ y=1 \end{array}\right\}$ (8) $\left.\begin{array}{l} x=7 \\ y=5 \end{array}\right\}$ (9) $\left.\begin{array}{l} x=2 \\ y=-10 \end{array}\right\}$

Exercise 64
(1) $\left.\begin{array}{l} x=2 \\ y=1 \end{array}\right\}$ (2) $\left.\begin{array}{l} x=4 \\ y=5 \end{array}\right\}$ (3) $\left.\begin{array}{l} x=3 \\ y=-3 \end{array}\right\}$ (4) $\left.\begin{array}{l} x=\frac{3}{4} \\ y=1 \end{array}\right\}$ (5) $\left.\begin{array}{l} x=-5 \\ y=4 \end{array}\right\}$ (6) $\left.\begin{array}{l} x=7 \\ y=10 \end{array}\right\}$

(7) $\left.\begin{array}{l} x=-2 \\ y=-5 \end{array}\right\}$ (8) $\left.\begin{array}{l} x=\frac{1}{5} \\ y=\frac{1}{4} \end{array}\right\}$ (9) $\left.\begin{array}{l} x=4 \\ y=-1 \end{array}\right\}$

Exercise 65
(1) $\left.\begin{array}{l} x=1 \\ y=3 \end{array}\right\}$ (2) $\left.\begin{array}{l} x=2 \\ y=5 \end{array}\right\}$ (3) $\left.\begin{array}{l} x=4 \\ y=1 \end{array}\right\}$ (4) $\left.\begin{array}{l} x=-3 \\ y=-2 \end{array}\right\}$ (5) $\left.\begin{array}{l} x=3 \\ y=1\frac{1}{2} \end{array}\right\}$ (6) $\left.\begin{array}{l} x=-1 \\ y=3 \end{array}\right\}$

(7) $\left.\begin{array}{l} x=5 \\ y=3 \end{array}\right\}$ (8) $\left.\begin{array}{l} x=1 \\ y=-7 \end{array}\right\}$ (9) $\left.\begin{array}{l} x=-1 \\ y=1 \end{array}\right\}$

Exercise 66
(1) $\left.\begin{array}{l} x=11 \\ y=5 \end{array}\right\}$ (2) $\left.\begin{array}{l} x=-4 \\ y=3 \end{array}\right\}$ (3) $\left.\begin{array}{l} x=8 \\ y=3 \end{array}\right\}$ (4) $\left.\begin{array}{l} x=1 \\ y=3 \end{array}\right\}$ (5) $\left.\begin{array}{l} x=\frac{1}{2} \\ y=-5 \end{array}\right\}$ (6) $\left.\begin{array}{l} x=\frac{2}{5} \\ y=3 \end{array}\right\}$

(7) $\left.\begin{array}{l} x=4 \\ y=3 \end{array}\right\}$ (8) $\left.\begin{array}{l} x=-1 \\ y=-2 \end{array}\right\}$ (9) $\left.\begin{array}{l} x=3 \\ y=-1 \end{array}\right\}$

Exercise 67
(1) 0, 2   (2) 0, -3 (3) 1, -4 (4) $-2, -2$ (5) 0, 10   (6) $-4, 5$
(7) 6, 6   (8) 0, -5 (9) 1, -1 (10) $-3, -6$ (11) 3, -2 (12) $-9, 7$
(13) 0, 5   (14) $-2, 7$ (15) $-3, -3$ (16) $-4, 2$ (17) 1, 10   (18) 7, 7
(19) 0, -6 (20) 7, -7 (21) $-8, -3$ (22) $-9, -11$ (23) $-1, 4$ (24) $-2, -2$

Exercise 68
(1) 0, -3 (2) 0, 12   (3) 0, 7   (4) 0, -1 (5) 0, -2 (6) $-2, -3$
(7) $-1, -4$ (8) $-2, -5$ (9) $-7, -8$ (10) $-1, -6$ (11) 1, 5   (12) 2, 4
(13) 3, 6   (14) 3, 4   (15) 5, 7   (16) 1, -3 (17) 7, -8 (18) $-2, 5$
(19) 4, -9 (20) -2, 3   (21) 3, -3 (22) 10, -10 (23) 7, -7 (24) 5, -5
(25) 13, -13 (26) $-3, -3$ (27) 1, 1   (28) 5, 5   (29) $-10, -10$ (30) 8, 8

Exercise 69

(1) $x^2 - 7x + 10 = 0$ (2) $x^2 - 2x - 3 = 0$ (3) $x^2 + 2x - 8 = 0$ (4) $x^2 + 6x + 5 = 0$
(5) $x^2 - 4 = 0$ (6) $x^2 + 12x + 36 = 0$ (7) $x^2 + 5x = 0$ (8) $x^2 - 6x = 0$
(9) $x^2 + 5x - 14 = 0$ (10) $x^2 - 7x + 12 = 0$ (11) $x^2 - x - 30 = 0$ (12) $x^2 - 8x = 0$
(13) $x^2 - 49 = 0$ (14) $x^2 + 2x - 15 = 0$ (15) $x^2 - 36 = 0$

Exercise 70

(1) $0, \frac{1}{2}$ (2) $0, -\frac{2}{5}$ (3) $-\frac{1}{5}, 2$ (4) $1\frac{1}{2}, -1$ (5) $4, -1\frac{1}{5}$ (6) $-\frac{1}{2}, -2$
(7) $0, \frac{2}{3}$ (8) $4\frac{1}{2}, 1$ (9) $\frac{4}{3}, -3$ (10) $1, 1\frac{1}{3}$ (11) $-\frac{1}{7}, -1$ (12) $\frac{3}{5}, -2$

Exercise 71

(1) $-3, -\frac{1}{2}$ (2) $-1, -\frac{2}{3}$ (3) $-3, -\frac{1}{5}$ (4) $2, \frac{1}{3}$ (5) $0, \frac{3}{5}$ (6) $1, 1\frac{1}{2}$ (7) $1, \frac{2}{5}$
(8) $-6, -\frac{1}{3}$ (9) $-4, -1\frac{1}{2}$ (10) $0, -4\frac{1}{2}$ (11) $4, 1\frac{1}{2}$ (12) $4, \frac{2}{5}$ (13) $-5, \frac{1}{2}$
(14) $4, -\frac{3}{5}$ (15) $0, 1\frac{1}{4}$

Exercise 72

(1) $3x^2 - 5x + 2 = 0$ (2) $5x^2 + 9x - 2 = 0$ (3) $5x^2 - 2x = 0$ (4) $3x^2 + 7x + 2 = 0$
(5) $5x^2 - 11x - 12 = 0$ (6) $7x^2 + x = 0$ (7) $3x^2 + 14x - 5 = 0$ (8) $5x^2 - 6x + 1 = 0$

Exercise 73

(1) $-2, -7$ (2) $2, 3$ (3) $12, -12$ (4) $-5, 6$ (5) $0, 1\frac{1}{2}$ (6) $2, -2$
(7) $3, -5$ (8) $1, -\frac{1}{2}$ (9) $-3, 8$

Exercise 74

(1) $a = 1, \ b = 6, \ c = 6$ (2) $a = 1, \ b = -9, \ c = 17$ (3) $a = 1, \ b = 3, \ c = -2$
(4) $a = 1, \ b = -10, \ c = 6$ (5) $a = 1, \ b = 4, \ c = 2$ (6) $a = 1, \ b = -5, \ c = 5$
(7) $a = 1, \ b = -1, \ c = -5$ (8) $a = 1, \ b = 3, \ c = -6$ (9) $a = 1, \ b = -3, \ c = -9$
(10) $a = 1, b = -6, c = 3$ (11) $a = 1, b = -4, c = -6$ (12) $a = 1, b = 3, c = 1$

Exercise 75

(1) $0.562, -3.562$ (2) $-0.646, 4.646$ (3) $-1.209, -5.792$ (4) $0.764,$
5.236 (5) $-0.303, 3.303$ (6) $0.542, -5.542$ (7) $1.193, -4.193$
(8) $-0.382, -2.618$ (9) $-1.702, 4.702$ (10) $0.697, 4.303$ (11) $-0.268, -3.732$
(12) $0.86, 8.14$

Exercise 76a

(1) $-0.634, -2.366$ (2) $-0.591, 2.257$ (3) $-0.358, 0.558$ (4) $0.721, -1.388$
(5) $0.293, 1.707$ (6) $-0.310, -1.290$

Exercise 76b

(1) ± 2.646 (2) Imaginary   (3) ± 2.236 (4) ± 3.873 (5) ± 2.449
(6) Imaginary   (7) ± 2 (8) ± 3.317 (9) ± 15 (10) ± 4.472 (11) Imaginary
(12) ± 1.732

Exercise 77

(1) 4 cm, 6 cm, 8 cm   (2) 7 cm, 2 cm, 7 cm   (3) 8 cm   (4) 5 cm, 12 cm, 13 cm; 30 cm^2
(5) $x = 3$; 7 cm   (6) 30°, 60°, 90°   (7) 40°, 70°, 70°; isosceles   (8) 10 cm; 30 cm
(9) $x = 3$; area = 42 cm^2 (10) $x = 10$; perimeter = 34 cm   (11) 4 units; 45 units^2
(12) 60°, 80°, 100°, 120°   (13) A 5 units, 6 units, 8 units; B: 5 units, 7 units, 7 units;
perimeter = 19 units   (14) 24 cm, 6 cm   (15) 5 cm, 8 cm; 40 cm^2 (16) 32 cm

Exercise 78

(1) (a) 6   (b) 5   (c) 76   (2) 7;   33 units   (3) (a) WX = 7 cm,   WZ = 3 cm,
YZ = 11 cm,   XY = 5 cm   (b) 27 cm^2 (4) (a) $x = 3$ units;   HG = 6 units,
GF = 6 units, FB = 8 units;   (b) 264 units^2 (c) 288 units^3

Exercise 79

(1) 4, 3   (2) 8, 3   (3) 2, −1   (4) −2, 3   (5) −1, −6   (6) 4, $2\frac{1}{2}$
(7) Maths book 10p, science book 8p   (8) Apple 4p, peach 11p   (9) 101°, 79°
(10) Litre of oil £1.20, sparking plug 30p   (11) Nut 2p, bolt 7p   (12) $A_1 = 8$ cm^2,
$A_2 = 12$ cm^2

Exercise 80

(1) (a) $x = 3$, $y = −2$ (b) 34 units   (2) (a) $x = 5$, $y = 2$;   AC = 12 units,
BD = 16 units   (b) 96 units^2 (3) $x = −4$, $y = 1\frac{1}{2}$; DE = DG = 7 cm, FE = FG = 12 cm
(4) $x = 6$, $y = 4$; 42 units

Exercise 81

(1) 7, 3   (2) 3, 6; −3, −6   (3) 10, −2   (4) 3, −1;   1, −3   (5) −5, −6
(6) 4,   7   (7) 6,   2   (8) 12,   10   (9) 8 m,   5 m   (10) 4 m × 4 m,   6 m × 6 m;
16 m^2,   36 m^2 (11) 5 cm   (12) −11, −9;   9,   11   (13) 5; −3   (14) 4,   6,   8; −8,
−6, −4   (15) 5 m, 15 m

Exercise 82

(1) (a) AC = 6.22 cm,   BC = 3.22 cm   (b) AB = 7 cm   (2) (a) $x = 7$;   XZ = 24 units,
YZ = 26 units   (b) 120 units^2 (3) (a) $x = 5.83$ cm,   PR = 23.32 cm,   QS = 5.66 cm
(b) 12 cm

Exercise 83

(1) (a) 37.68   (b) 29.52   (2) (a) 20   (b) 4.44   (3) (a) 84   (b) 92.07
(4) (a) 50   (b) 42   (5) (a) 502.4   (b) 14.95   (6) (a) 31.4   (b) 13.13
(7) (a) 25   (b) 16   (8) (a) 27   (b) 10.25   (9) (a) 45   (b) 12.04
(10) (a) 12   (b) 4.47   (11) (a) 30   (b) 35.06   (12) (a) 175.84 (b) 158.3
(13) (a) 192   (b) 92   (14) (a) 40   (b) 7.55   (15) (a) 45   (b) 100
(16) (a) 314   (b) 162.78   (17) (a) 13   (b) 5.77   (18) (a) 33.49   (b) 7.23
(19) (a) 188.4   (b) 46.89   (20) (a) 12   (b) 15.85   (21) (a) 12.56   (b) 1.01
(22) (a) 5.57   (b) 3.71   (23) (a) 295.6   (b) 202.94   (24) (a) $\frac{1}{2}$, −4   (b) 5.45, 0.55

Exercise 84

(1) $r = \dfrac{C}{2\pi}$ (2) $b = \dfrac{2A}{h}$ (3) $l = \dfrac{A}{\pi r}$ (4) $T = \dfrac{D}{S}$ (5) $\sin C = \dfrac{2A}{ab}$ (6) $s = \dfrac{(v^2 - u^2)}{2f}$

(7) $u = \sqrt{v^2 - 2fs}$ (8) $T = \dfrac{100I}{PR}$ (9) $z = \sqrt{x^2 - y^2}$ (10) $r = \sqrt{\dfrac{A}{\pi}}$ (11) $g = \dfrac{4\pi^2 l}{T^2}$

(12) $R = \sqrt{\dfrac{A + \pi r^2}{\pi}}$

Exercise 85

(1) $H = \dfrac{V}{LB}$; 5   (2) $D = ST$; 178.75   (3) $A = \dfrac{3V}{h}$; 24   (4) $f = \dfrac{v - u}{t}$; 8.5

(5) $y = \sqrt{x^2 - z^2}$; 7   (6) $h = \dfrac{2A}{(x+y)}$; 6.2   (7) $P = \dfrac{100I}{RT}$; 600   (8) $s = \dfrac{v^2 - u^2}{2f}$; 2.61

(9) $l = \dfrac{gT^2}{4\pi^2}$; 154.51   (10) $\cos A = \dfrac{b^2 + c^2 - a^2}{2bc}$; $73°24'$ (11) $r = \sqrt[3]{\dfrac{3V}{4\pi}}$; 4.92

(12) $r = \sqrt{\dfrac{\pi R^2 - A}{\pi}}$; 8.96

Exercise 86
(1) y^5 (2) 5^7 (3) x^7 (4) $12y^4$ (5) a^3 (6) 2^3 (7) $x^3 y^{-3}$ (8) 3^{12} (9) 5^6
(10) 2^5 (11) 4^7 (12) 5^3 (13) x^4 (14) 2^7 (15) 2^{-7} (16) $m^0 = 1$ (17) 4^{-2}
(18) $\dfrac{x^2}{2}$ (19) $2^0 = 1$ (20) 5^6 (21) a^7 (22) $2^0 = 1$ (23) $\frac{1}{3} = 3^{-1}$ (24) l^2
(25) 5^4 (26) $27x^{12}$ (27) 2^{12} (28) 36   (29) $x^{10} y^{15}$ (30) $8a^3 b^6$ (31) $\dfrac{16x^8}{81y^4}$
(32) 1   (33) 2^{10} (34) $4x^{10} y^{12}$ (35) 4^4 (36) 5^2

Exercise 87
(1) 1   (2) 3   (3) 2   (4) 2   (5) $\frac{1}{8}$ (6) $\frac{1}{8}$ (7) $\frac{1}{9}$ (8) $\frac{1}{25}$ (9) 2   (10) $\frac{16}{9}$
(11) $\frac{1}{64}$ (12) 1   (13) 4   (14) 8   (15) 27   (16) $\frac{1}{32}$ (17) $\frac{1}{2}$ (18) $\frac{1}{9}$ (19) $\frac{1}{3}$
(20) $\frac{1}{5}$ (21) $\frac{4}{5}$ (22) $\frac{2}{3}$ (23) $\frac{1}{8}$ (24) 2   (25) 16   (26) $2x^2$ (27) $\frac{1}{5}$ (28) 128
(29) 50   (30) $1\frac{1}{2}$ (31) 3^{3x} (32) 4^{2x}

Exercise 88
(1) 3   (2) 2   (3) $\frac{1}{3}$ (4) -2 (5) 3   (6) 2   (7) $-2\frac{1}{2}$ (8) $-1\frac{2}{3}$ (9) $\frac{1}{4}$
(10) 2   (11) 3   (12) 1

Exercise 89
(1) 5.2×10^4 (2) 4.95×10^{-6} (3) 6.5×10^{10} (4) 8.15×10^{-3} (5) 9.9×10^8
(6) 4.0×10^{-10} (7) 2.5×10^6 (8) 7.3×10^{-3} (9) 5.12×10^{14} (10) 8.5×10^{-13}

Exercise 90
(1) 1200   (2) 0.0000626   (3) 98 000 000 000   (4) 0.037   (5) 5 200 000
(6) 0.000 000 001 1   (7) 0.000 041 7   (8) 100 000   (9) 8 950 000 000 000
(10) 0.000 000 023 5

Exercise 91
(1) Mercury 5.8×10^7 Venus 1.08×10^8 Earth 1.5×10^8 Mars 2.28×10^8
Jupiter 7.78×10^8 Saturn 1.427×10^9 Uranus 2.87×10^9 Neptune 4.497×10^9
Pluto 5.91×10^9 (2) Japan 105 millions India 550 millions China 732 millions
UK 55 millions USA 203 millions USSR 242 millions   (3) (a) 0.000 000 000 002 g
(b) 0.000 000 000 000 001 m   (c) 8.0×10^{-6} mm   (d) 1.0×10^{-3} cm
(4) (a) 4 700 000 000 years   (b) 5.0×10^9 years   (c) 50 000 000 000 000 cells
(d) 1.0×10^{15} vibrations per second

Exercise 92
(1) 4.48×10^9 (2) 1.75×10^6 (3) 1.164×10^3 (4) 1.611×10^{-13}
(5) 6.875×10^{-12} (6) 6.768×10^8 (7) (a) (i) 3.879×10^{15} km   (ii) 4.058
$\times 10^{13}$ km   (iii) 1.514×10^{16} km   (iv) 1.892×10^{19} km   (b) (i) 5.97 light years

(ii) 8.7 light years   (iii) 107 light years   (iv) 5000 light years
(8) (a) 7.0×10^{-5} cm   (b) 3.5×10^{-5} cm

Exercise 93

(1) $x = -3$ (2) $\dfrac{a+b}{ab}$ (3) -5 (4) $2\frac{1}{2}$ (5) 1.6×10^{-3} (6) $x = 3$ (7) $2A - B$

(8) 0.0000215 (9) 18   (10) None of these   (11) 5; 4   (12) $\dfrac{f-a}{g}$ (13) 3140

(14) $x^2 + 4x - 12$ (15) $x = 3$ (16) 32   (17) $2k^3$ (18) $\frac{1}{2}$ (19) $(x+6)(x-2)$
(20) b^2c^6 (21) $\frac{1}{16}$ (22) $4x^8$ (23) $4p^3$ (24) -2 (25) 4   (26) $\frac{3}{2}$ (27) $2a$
(28) $2x^2 - x - 6$ (29) 2   (30) 1.25

Exercise 94

(1) $f = \dfrac{2(s - ut)}{t^2}$ (2) $x = 2$, $y = 7$ (3) $x = 12$ (4) (a) $10x - 5x^3$ (b) $10 + 3x - x^2$

(c) $2x(x^2 - 5x + 3)$ (d) $(x-3)(x-4)$ (e) $x(y - z + x)$ (5) (a) $x = 5$ (b) $x = 2$
(c) $x = 1$ (d) $x = 3$, $y = 2$ (6) (a) 10   (b) 0   (c) 52   (d) 2   (e) $\frac{1}{4}$ (7) $4x$ (8) -5

(9) (a) $2x$ (b) $x^2 - 1$ (c) -1 (d) $x = 3$ (e) $x = -\dfrac{(1+y)}{(1-y)}$ (10) 18   (11) $x = -1$,

$y = 1$ (12) 20   (13) $y = \dfrac{x - 13}{5}$ (14) $a = 3$, $b = 1$ (15) $x = 2$ (16) $\dfrac{3}{x(3 + 2x)}$

(17) $8x^2 - 14x - 15$ (18) (a) -1 (b) 8   (c) 75   (19) (a) $7x$ (b) $3x$ (c) x^2
(20) (a) $x = 4$ (b) $y = 1$ (c) $y = 1$ (d) 2   (21) 12   (22) $2\frac{1}{2}$ (23) 16   (24) $x - 7y$
(25) (a) $x = 9$ (b) 2^8 (26) -12 (27) 0   (28) $3x - 4$ (29) (a) $3ac(b - 3d)$
(b) $(a+1)(b-c)$ (30) $a = 3$, $b = 0$ (31) (a) $12a$ (b) $-6a$ (c) 3   (d) $-27a^2$
(32) $1, -\frac{3}{5}$ (33) (a) 44   (b) 154   (c) 396   (d) 462   (34) (a) 36   (b) 8
(35) $x = 2$ (36) 21   (37) $x = 5$, $y = 2$ (38) $x = \pm 4$ (39) $x = 12$ (40) $36°$

(41) 616 cm^2 (42) 4080   (43) $\dfrac{A - 2\pi r^2}{2\pi r}$ (44) $100°$ (45) $7, -2$

(46) $2a(4a - 3b + 1)$ (47) 21   (48) 3, 5   (49) $36°$, $36°$, $108°$ (50) $x = 4$, $y = 1$

(51) $x = 2$ (52) $\dfrac{v - u}{f}$ (53) (a) $4a$ (b) $3a$ (c) $4a^2$ (d) 4   (e) $4a^2$ (54) (a) 2

(b) -4 (c) 4   (d) -3 (e) 1   (f) 6   (55) (a) $11a - 10b$ (b) $a^2 - 5a + 6$
(56) $x = 3$, $y = 5$ (57) $5\frac{1}{2}$ (58) (a) $-1\frac{1}{2}$ (b) $-37\frac{1}{2}$ (c) 18   (d) $-3\frac{1}{3}$ (e) 16

(59) (a) $6 + 3x$ (b) $x - y$ (c) $4x$ (d) $\dfrac{x^2}{3y}$ (60) (a) $x = 8$ (b) $y = 3$ (c) $x = 1\frac{1}{2}$, $y = 3$

(d) $x = 10$

Exercise 95

(1) (a) (i) 42   (ii) $\frac{2}{3}$ (iii) 16   (b) (i) 9   (ii) 5   (iii) $1\frac{1}{25}$ (c) $x = 20$ m,
$y = 30$ m   (2) (a) (i) $x = 2$ (ii) $x = 12$ (b) (i) $2(a+2)$ (ii) $(2+a)(x+2y)$
(iii) $(2x-1)(2x+1)$ (c) $-0.21, -4.79$ (3) (a) 1.467×10^{25} (b) (i) $x = 1\frac{1}{4}$
(ii) $5, -5$ (iii) $5.58, -3.58$ (4) (a) (i) $(x-y)(x+y)$ (ii) 7372   (b) 16.9

(c) $1.39, -0.72$ (5) (a) 6.9×10^4 (b) 2.0×10^3 (6) (a) $\dfrac{8x}{15}$ (b) $x = 3$, $y = 7$

(c) (i) ± 4 (ii) $\frac{1}{16}$ (iii) 16   (7) (a) $k = 16$ (b) $P = \dfrac{FW}{(2Wk - F)}$ (c) $P = 3$

(8) (a) $n = 2$ (b) $n = 0$ (c) $n = -\frac{1}{3}$ (d) $n = 1$ (9) (a) 4.56, 0.44 (b) (i) $x + y = 210$
(ii) $40x + 20y = 6000$ (iii) 90 adult customers   (10) (a) $2x^2 + xy - 2y^2$
(b) $\dfrac{-1}{(x-2)(x-3)}$ (c) $0.19, -5.19$ (d) $A = P(1+R)^2$ (11) (a) $x = 15$, $y = 8$
(b) (i) $(x^2 + 6x + 9)$ square units; $(3x^2 + 3x)$ square units   (ii) $x = 3$ (iii) Square:
6 units × 6 units;   Rectangle: 9 units × 4 units   (12) (a) 2.59   (b) $6x^2 - 5xy - 4y^2$
(c) $k = 3$, $x = 27$ (d) $\frac{1}{2}, -2$ (13) (a) $2x$ miles   (b) $12x$ miles   (c) $14x$ miles
(d) $14x = 42$, $x = 3$ (e) 6 miles   (f) 36 miles   (g) $8\frac{2}{5}$ miles per hour   (14) (a) (i) 1
(ii) 5   (iii) 7   (b) $4, -1\frac{1}{2}$ (c) $1, -4$ (d) $y = 36$ (15) (a) (i) 64   (ii) 1   (iii) ± 2
(b) $\dfrac{by}{x}$ metres   (c) $a = 5, b = 6$ (d) $x = 4$ (16) (a) $n = 2$ (b) $n = 0$ (c) $n = -1$
(d) $n = \frac{1}{2}$ (17) (a) $x = 3$ (b) $x = 10, y = 2$ (c) (i) 11   (ii) 8   (iii) 2   (d) (i) 270
(ii) $n = 1$ (18) (a) $4851\,\mathrm{cm}^2$ (b) 11 cm   (c) $10\frac{1}{2}$ cm   (19) (a) (i) $(4-y)(x-6)$
(ii) $(3x-2)(x-1)$ (b) (i) 3   (ii) $\frac{1}{4}$ (iii) 8   (c) $OC = 3\,\mathrm{cm}$ (20) (a) $523.33\,\mathrm{m}^3$
(b) $272.94\,\mathrm{m}^2$ (21) (a) (i) $xy - y^2$ (ii) $x = 9$ cm, $y = 6$ cm   (iii) $\frac{2}{3}$
(b) (i) $c = \dfrac{100w + mp}{100}$ (ii) $p = \dfrac{100(c-w)}{m}$ (22) (a) $x = -2$, $y = 3\frac{1}{2}$ (b) $x = 0$,
$x = 4$ (c) (i) 6 m   (ii) $(x-4)$ m   (iii) $(6x - 24)\,\mathrm{m}^2, x = 14$ m

LINKED

027573

S. KATHARINE'S COLLEGE
LIBRARY